R. B. Woodward and R. Hoffmann

The Conservation of Orbital Symmetry

R. B. Woodward and R. Hoffmann

The Conservation of Orbital Symmetry

Verlag Chemie International

Deerfield Beach, Florida

With 45 Figures

Prof. Robert Burns Woodward
Department of Chemistry, Harvard University
12 Oxford Street
Cambridge, Mass. 021 38 (USA)
Prof. Roald Hoffmann
Department of Chemistry, Cornell University
Ithaca, N.Y. 14850 (USA)

First Edition 1970
First Printing January 1970
Second Printing May 1970
Third Printing June 1981

Library of Congress Catalog Card Number 79-103 636

ISBN 0-89573-109-6 (Deerfield Beach)
ISBN 3-527-25324-6 (Weinheim)

Printed in the United States of America

Contents

Errata

Page 80, formula *(173)*: the asterisk should be placed at the CH_2 group of the five-membered ring, as in *(174)*.

Page 83, line 1: replace calicenes by sesquifulvalenes.

Page 86, footnote [100], last reference: read *88*, 2880 (1966).

Page 103, line 9: read *(273)* instead of *(278)*.

Page 120, footnote [142], first reference: read *89*, 5503 (1967).

Page 125, figure 36: the two formulae must be interchanged.

Page 130, footnote [177]: read *1966*, 6393.

Page 142, footnote [185]: add *E. E. van Tamelen, R. S. Dewey,* and *R. J. Timmons,* ibid. *83,* 3725 (1961).

Page 143, footnote [187]: add *K. Schaffner,* personal communication.

Page 176, line 11: read *Laidler* [249a] and *Shuler* [249b].

Page 176, footnote [249]: insert a) in front of the first reference and add b) *K. E. Shuler,* J. chem. Physics *21*, 624 (1953).

1. Introduction

Of the various ways in which the phenomena of chemical bonding have been treated from the theoretical point of view, the molecular orbital method may fairly be deemed to have been the most fruitful in the hands of the organic chemist, and most adaptable to his needs. None the less, the method, with a few conspicuous exceptions, has been used mainly in the study of the static properties of molecules — in ground and excited states — and only rarely have its potentialities been explored in relation to reacting systems.

In 1965, in a series of preliminary communications[1-3], we laid down some fundamental bases for the theoretical treatment of all concerted reactions. The history of the genesis of these ideas has been described elsewhere[4]. The basic principle enunciated was that reactions occur readily when there is congruence between orbital symmetry characteristics of reactants and products, and only with difficulty when that congruence does not obtain — or to put it more succinctly, *orbital symmetry is conserved in concerted reactions*. This principle has met with widespread interest; the applications made of it, the tests which it has survived, and the corollary predictions which have been verifed are already impressive. In this paper we develop our views at some length, survey some of the developments — in our hands and those of others — of the three years just past, and make some new projections.

---

[1] *R. B. Woodward* and *Roald Hoffmann*, J. Amer. chem. Soc. *87*, 395 (1965).
[2] *Roald Hoffmann* and *R. B. Woodward*, J. Amer. chem. Soc. *87*, 2046 (1965).
[3] *R. B. Woodward* and *Roald Hoffmann*, J. Amer. chem. Soc. *87*, 2511 (1965).
[4] *R. B. Woodward:* Aromaticity. Special Publication No. 21. The Chemical Society, London 1967, p. 217.

2. Orbitals and Bonding

It is worth while to review the elementary aspects of the molecular orbital theory of bonding[5]. Molecular orbitals are constructed as combinations of atomic orbitals, and are then populated by electron pairs. When two equivalent atomic orbitals, χ_1 and χ_2, combine, they always yield a bonding combination and a corresponding antibonding orbital *(1)*.

χ_1 — χ_2

(1)

The bonding combination is characterized by positive overlap, and by concentration of electron density in the region between the nuclei. By contrast, the antibonding combination exhibits negative overlap, and a nodal surface in the region between the nuclei. When χ_1 and χ_2 are *s* orbitals, the bonding combination is $\chi_1+\chi_2$, and the antibonding one $\chi_1-\chi_2$.

$\chi_1 + \chi_2$

$\chi_1 - \chi_2$

When χ_1 and χ_2 are *p* orbitals interacting in a σ manner, and oriented as shown in *(2)* the bonding combination is again $\chi_1+\chi_2$ and the antibonding combination $\chi_1-\chi_2$.

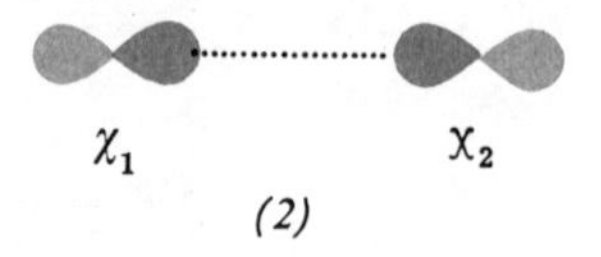

(2)

Throughout this paper, the phases of wave-functions are color-coded: positive = blue, negative = green. When phase relationships are not relevant, the orbitals are presented in solid gray.

---

[5] In addition to the classical text by *C. A. Coulson* (Valence. 2nd ed., Oxford University Press, London 1961), we recommend: *C. A. Coulson* and *E. T. Stewart* in *S. Patai:* The Chemistry of Alkenes. Wiley-Intersience, New York 1964, *E. Heilbronner* and *H. Bock:* Das HMO-Modell und seine Anwendung. Verlag Chemie, Weinheim 1968, and *R. S. Mulliken*, Science *157*, 13 (1967); Angew. Chem. *79*, 541 (1967).

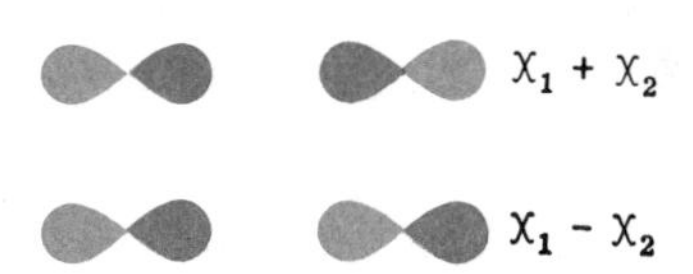

It is important to realize, however, that were the basis orbitals initially arbitrarily oriented in some other fashion, such as *(3)*, then, since $\chi'_2 = -\chi_2$, the bonding

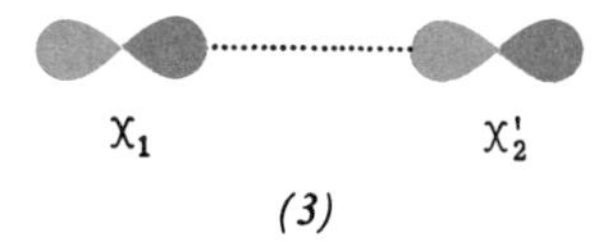

(3)

combination would be $\chi_1 - \chi'_2$, and the antibonding one $\chi_1 + \chi'_2$. It should also be kept in mind that multiplying an entire wave function by -1 does not affect its energy. Thus, overlap of minus with minus lobes is precisely equivalent to plus with plus lobes, and $-\chi_1 - \chi_2$ is the *same* bonding orbital as $\chi_1 + \chi_2$.

The description of σ bonds in hydrocarbons is simple. Each formal chemical bond engenders a σ and a σ^* orbital. The C-H and C-C cases are:

The molecular orbitals are represented in our drawings as the overlap of two hybrids of unspecified hybridization. It should be emphasized that this is only intended as an artistic mnemonic device: the only essential features of a σ orbital are that it is approximately cylindrically symmetrical around the bond axis, that it concentrates electron density in the region between the nuclei, and that there is no nodal plane between the atoms.

Our simple picture of bonding in, say, cyclobutane, shows four C-C σ levels, and eight C-H σ levels, each with a matching σ^* level *(4)*. The carbon 1*s* orbitals are

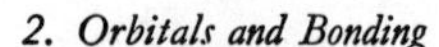

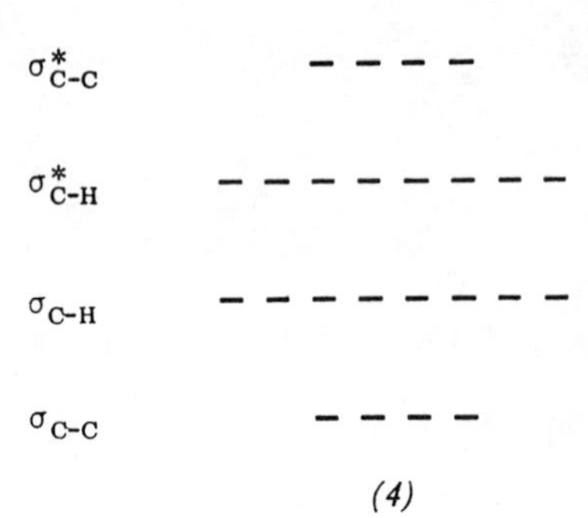

(4)

not considered. Each of the σ levels is occupied by two electrons. Spectroscopic studies indicate that the gap between occupied and unoccupied levels must be of the order of 10 electron volts.

The orbitals we have constructed are semi-localized. They are delocalized over only two atoms. Such orbitals are satisfactory for an analysis of some properties of the molecules — those which depend on all of the occupied molecular orbitals: bond lengths, energies, dipole moments. They are not the proper molecular orbitals of the molecule. The latter are completely delocalized, subject to the full symmetry of the molecule. For a discussion of physical properties depending on one or two specific molecular orbitals, such as spectra or ionization, it is absolutely necessary to construct these equivalent delocalized orbitals. The mechanics of delocalization will be described in Section 3.

In addition to σ bonds, the molecules of organic chemistry contain delocalized π orbitals. Thus, the electronic structure of ethylene is described as follows: There are four σ C-H bonds and a σ C-C bond *(5)*. Each generates a σ and a σ^* level, and

(5) *(6)*

five pairs of electrons are placed in the σ levels. There remain two electrons and two atomic p orbitals perpendicular to the plane of the molecule *(6)*. These combine to give π and π^* orbitals, differentiated by the absence or presence of a node between the atoms (Figure 1).

There are two independent symmetry operations which may be used to classify these orbitals; the mirror plane m, perpendicular to the molecular plane and bisecting the molecule, and the two-fold rotation axis C_2, passing through the center of the carbon-carbon bond. It should be noted carefully that the symmetry proper-

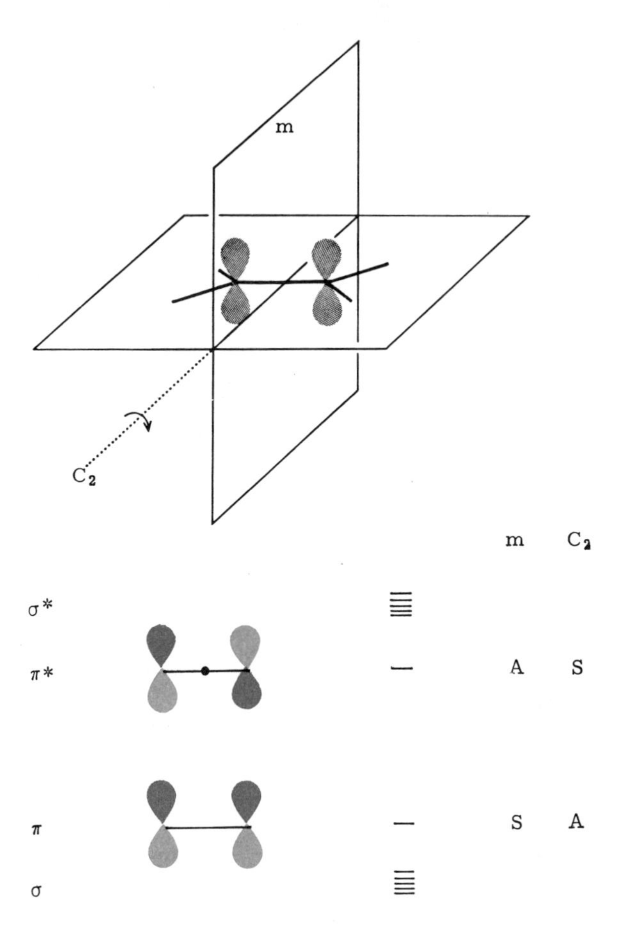

Figure 1. Molecular orbitals of ethylene and symmetry properties of the π and π^* orbitals. A = antisymmetric, S = symmetric. The horizontal bars indicate the relative orbital energies.

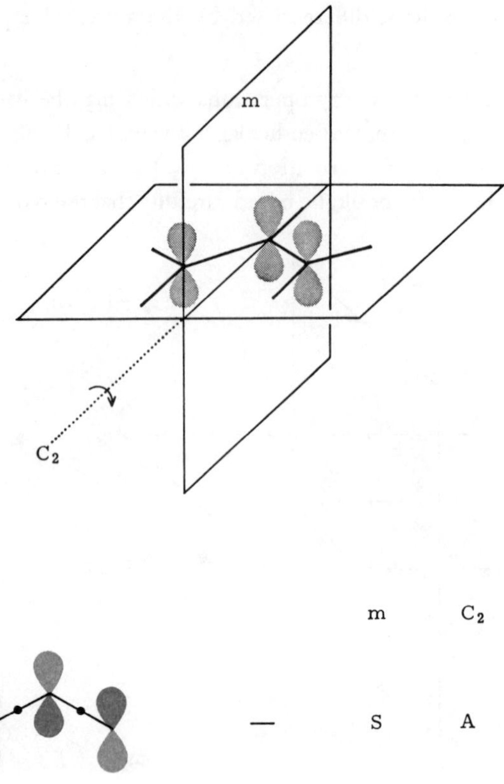

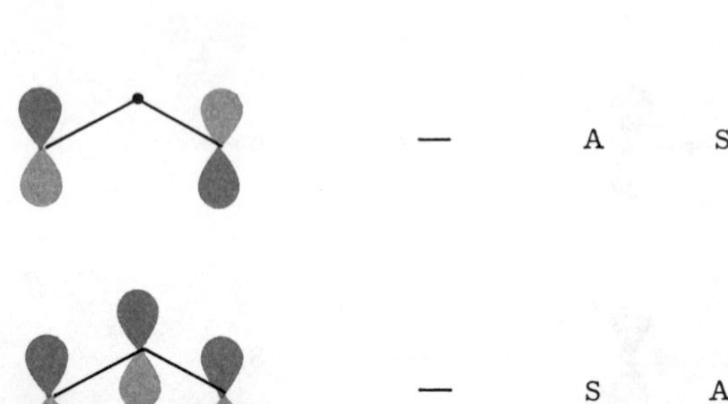

Figure 2. π molecular orbitals of the allyl system. A = antisymmetric, S = symmetric.

Figure 3. π molecular orbitals of *s-cis*-butadiene. A = antisymmetric, S = symmetric.

⟶

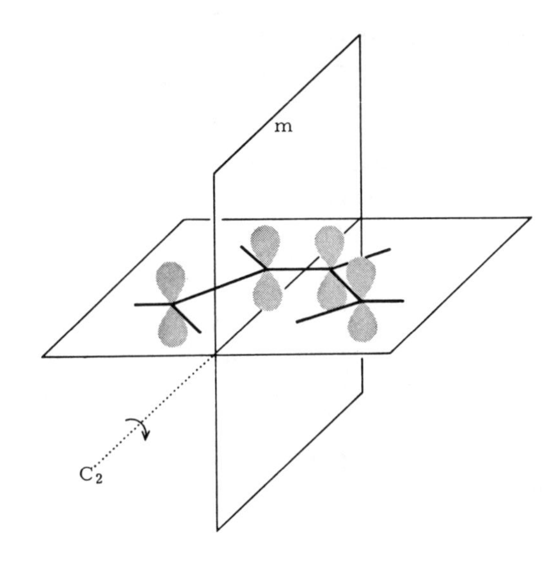

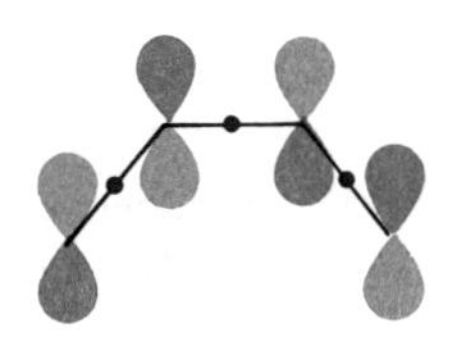

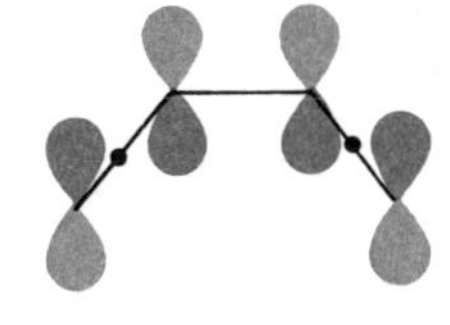

— S A

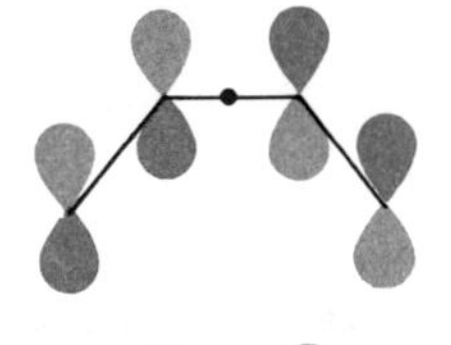

— A S

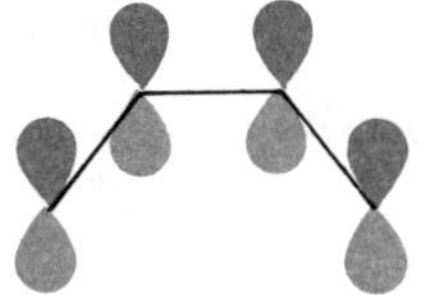

— S A

ties of the ethylene orbitals under each of the above operations are precisely opposite. Thus, the π orbital is symmetric (S) with respect to the mirror plane m, and antisymmetric (A) with respect to the rotation axis C_2.

The overlap between the two $2p_z$ orbitals is significantly less than that involved in σ interactions, and so the π bond is weaker than a σ bond, and the π and π^* levels are raised and lowered, respectively, from the sea of σ and σ^* levels (Figure 1).

The π molecular orbitals of the three-orbital allyl system are shown in Figure 2. Their nodal structure should be carefully noted. By virtue of their character as π orbitals, they all are antisymmetric under reflection in the plane of the allyl system. The lowest orbital, doubly occupied in the allyl cation, has no additional nodes. The middle, nonbonding orbital, which is singly occupied in the allyl radical, and doubly occupied in the anion, has a single nodal plane which precludes any contribution of the $2p$ orbital at the central carbon atom. The orbital of highest energy has two nodes.

The molecular orbitals of the four-orbital butadiene system are shown in Figure 3 for an *s-cis* arrangement[6,7]. Note once again the nodal structure and the alternating symmetry properties. The correlation of higher energy with an increasing number of nodes is not an accident, but rather a general consequence of either classical or quantum mechanics. The envelopes of polyene orbitals coincide with the curve of the wave function of a particle in a one-dimensional box *(7)*. The lowest orbital has no nodes, the next higher one has one node, the next two, and so on until the highest orbital has the maximum number of nodes possible. The general expression for the kth molecular orbital of a polyene or polyenyl system with n carbon atoms is

$$\Psi_k = \sum_{i=1}^{n} C_{ki}\, \Phi_i,$$

[6] Throughout this paper molecular orbitals are symbolized in terms of the atomic orbitals whose interaction gives the actual molecular orbital; since we are in general interested only in nodal properties, we ignore the fact that the coefficients – and thus the relative sizes – of the resultant atomic orbital contributions are not all identical.

[7] The electronic structure of polyenes is perhaps the most highly developed branch of semi-empirical molecular orbital theory. A very good survey of the field is given in *L. Salem:* The Molecular Orbital Theory of Conjugated Systems. Benjamin, New York 1966. See also *A. Streitwieser:* Molecular Orbital Theory for Organic Chemists. Wiley, New York 1961.

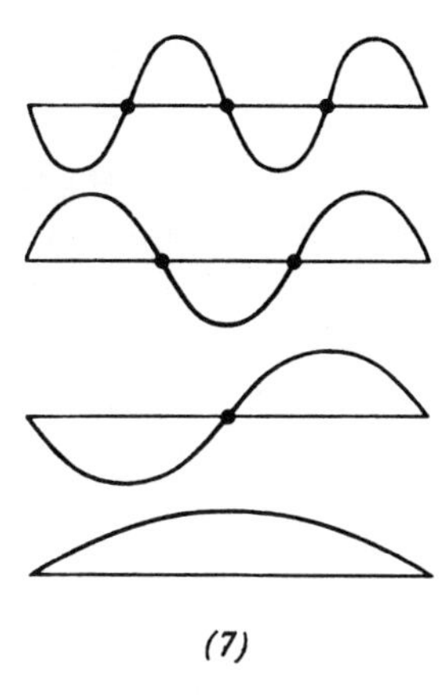

(7)

where the Φ_i are the atomic orbitals numbered consecutively from one end. The coefficients are given by the expression[7]

$$C_{ki} = \sqrt{\frac{2}{n+1}}\ \sin\frac{\pi ki}{n+1}$$

The orbitals alternate in symmetry with increasing energy.

If n is even, there are $n/2$ bonding π orbitals and $n/2$ antibonding. If n is odd, there are $(n-1)/2$ bonding, $(n-1)/2$ antibonding, and one nonbonding orbital.

A final point of much importance is that no molecular orbital may be at the same time symmetric and antisymmetric with respect to any existing molecular symme-

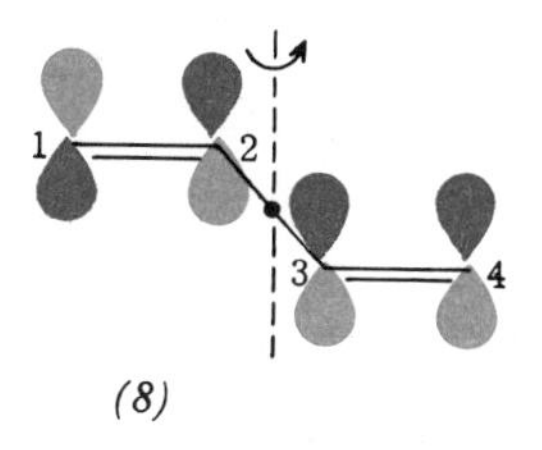

(8)

try element. Thus, the orbital *(8)* is symmetric under rotation by 180° if atoms 2 and 3 are viewed, but antisymmetric if 1 and 4 are considered; it is not an acceptable molecular orbital of butadiene.

3. Correlation Diagrams

The united atom-separated atoms diatomic correlation diagrams first drawn in the early nineteen-thirties by *Hund* and *Mulliken* have an important place in theoretical chemistry[8]. In constructing such a diagram, one imagined the process of two atoms approaching each other from infinity. The energy levels of the separated atoms were placed in approximate order of energy on one side of the diagram. One then imagined the approach of the atoms through the physically realistic molecu-

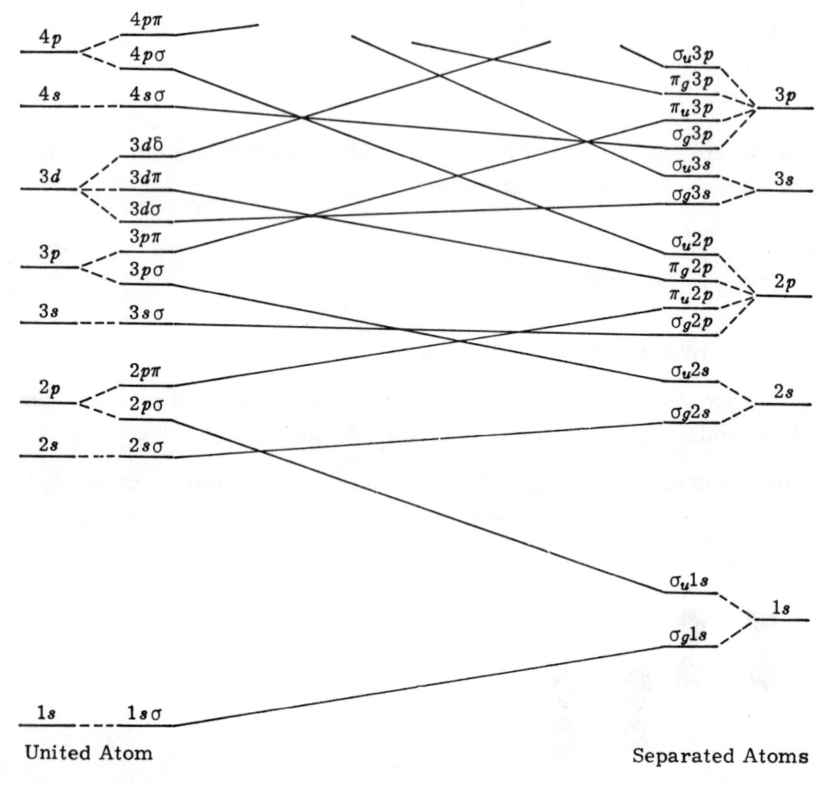

Figure 4. A typical atomic correlation diagram.

---

[8] *F. Hund*, Z. Phys. *40*, 742 (1927); *42*, 93 (1927); *51*, 759 (1928); *R. S. Mulliken*, Phys. Rev. *32*, 186 (1928); Rev. Mod. Phys. *4*, 1 (1932). See also *G. Herzberg:* The Electronic Structure of Diatomic Molecules. 2nd edition, Van Nostrand, Princeton 1950.

lar region into the physically impossible process of nuclear coalescence. The energy levels of the resulting united atom were once again known. They were placed on the other side of the diagram. One then proceeded to classify the initial separated and the final united atom orbitals with respect to the symmetry maintained throughout the hypothetical reaction. Levels of like symmetry were connected, paying due attention to the quantum mechanical noncrossing rule — that is, only levels of unlike symmetry are allowed to cross (Figure 4).

In this way, from the relatively well-known level structures of the separated atoms and the united atom valuable information was obtained about the level structure of the intermediate region corresponding to the molecule. It was this kind of diagram which provided a rationalization for the existence of the oxygen molecule as a ground-state triplet.

In an exactly analogous manner a correlation diagram may be drawn for a concerted reaction such as cycloaddition. On one side one writes down the approximately known energy levels of the reactants, on the other side those of the product. Assuming a certain geometry of approach one can classify levels on both sides with respect to the symmetry maintained throughout the approach, and then connect levels of like symmetry. Such a molecular correlation diagram yields valuable information about the intermediate region, which represents in this case the transition state for the reaction.

We would like to illustrate in some detail the construction of a molecular correlation diagram. The first example we choose is the maximum-symmetry approach of two ethylene molecules, leading to cyclobutane (Figure 5). As usual in theoretical discussions, maximum insight into the problem at hand is gained by simplifying

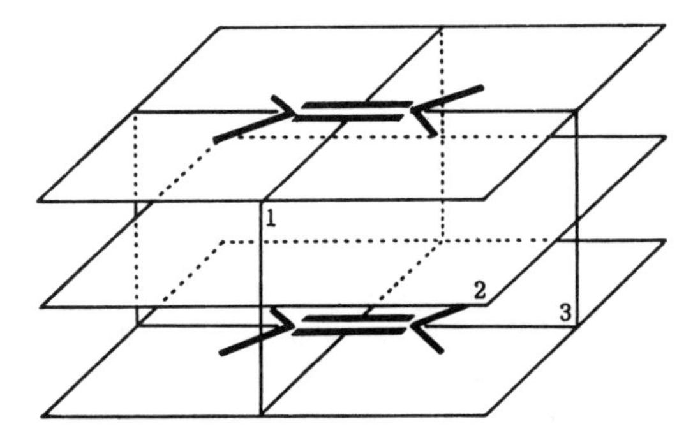

Figure 5. Parallel approach of two ethylene molecules.

the case as much as possible, while maintaining the essential physical features. In this instance we treat in the correlation diagram *only four* orbitals—the four π orbitals of the two ethylenes. In the course of the reaction these four π orbitals are transformed into four σ orbitals of cyclobutane. We may safely omit the C-H and the C-C σ bonds of the ethylene skeleton from the correlation diagram because, while they undergo hybridization changes in the course of the reaction, their number, their approximate positions in energy, and in particular their symmetry properties are unchanged.

The first step in the construction of a correlation diagram involves isolating the essential bonds and placing them at their approximate energy levels in reactants and

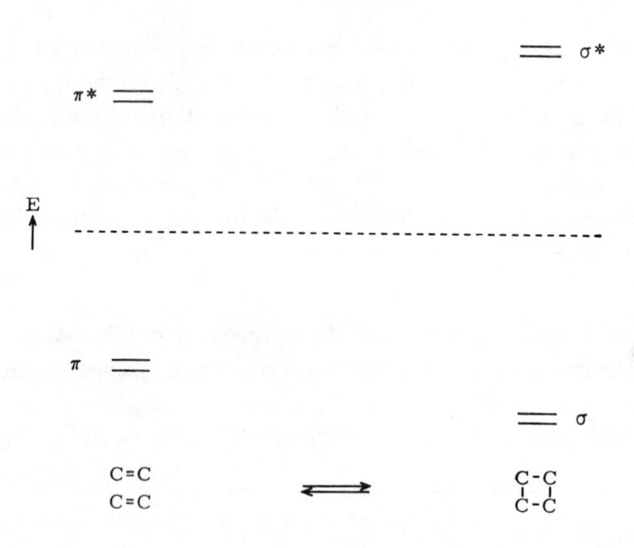

Figure 6. Energy levels of the orbitals essential for the formation of cyclobutane from two ethylene molecules in the geometry shown in Figure 5.

products; the result is shown for the case at hand in Figure 6, in which the dashed horizontal line is the nonbonding level— approximately the energy of an electron in a free carbon $2p$ orbital. We have separated σ and σ^* by an energy greater than that between π and π^*; although there is little reason to question this assignment, it is important to emphasize that it is in no way essential to the subsequent argument. To assign an order of magnitude to the vertical energy scale we may take the distance between π and π^* as approximately 5 electron volts[7].

In the next step the proper molecular orbitals for the reactants and products are written down. A digression on constructing molecular orbitals for interacting sys-

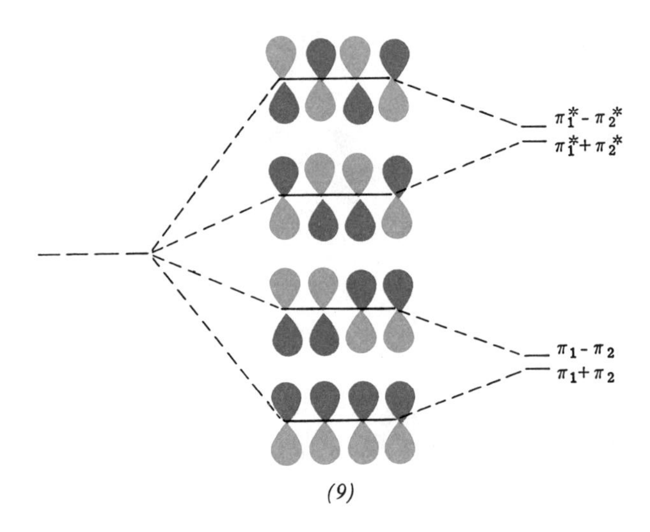

(9)

tems is in order at this point. Consider an *s-trans*-butadiene. One way to derive the four molecular orbitals of butadiene is to allow all of the four atomic orbitals to interact as illustrated at left in *(9)*.

Yet another way is to view butadiene as arising from the interaction of two semilocalized double bonds. Consider the bonding π orbitals of the two ethylenes. While π_1 and π_2 are entirely satisfactory for a description of the isolated double bonds, they are not the proper combinations to use as molecular orbitals for butadiene. Molecular orbitals must be symmetric or antisymmetric with respect to any molecular symmetry element which may be present. In the case at hand the crucial symmetry operation is a 180° rotation around the two-fold axis. The obvious combinations to use, and ones which do satisfy the symmetry conditions, are $\pi_1 \pm \pi_2$ *(10)*. These, of course, turn out to be topologically identical with the two lowest-energy butadiene orbitals which were constructed by considering direct interaction of all four atomic orbitals.

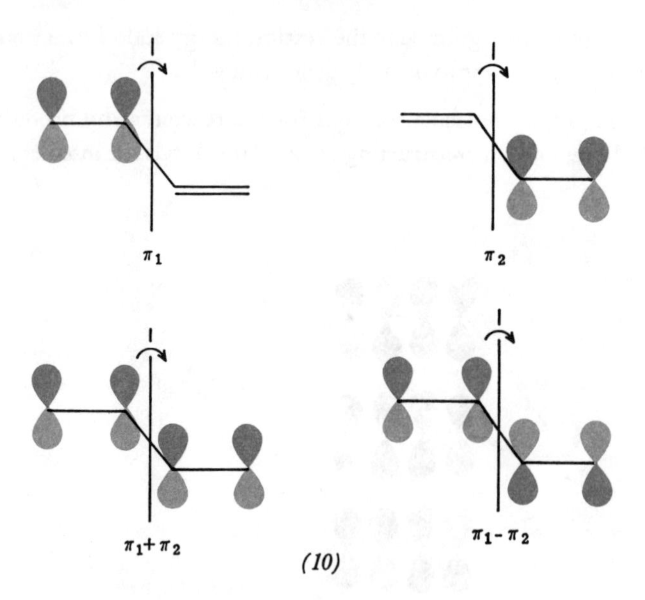

(10)

We are now prepared to treat the analogous problem of the molecular orbitals of two ethylenes approaching each other. Drawing orbital cross-sections in plane 3

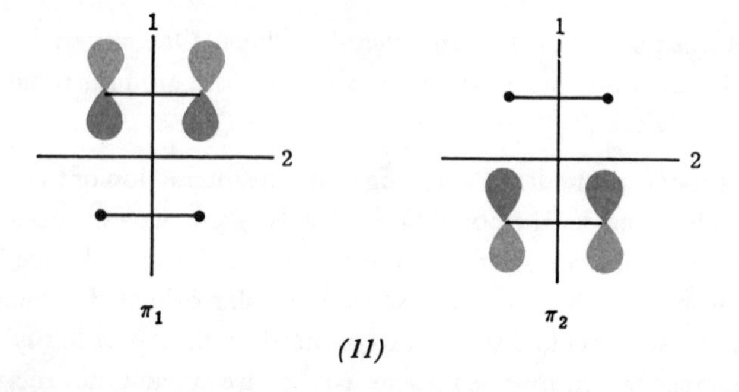

(11)

(*cf.* Figure 5), diagram *(11)* represents localized π bonds of the two ethylenes. These are not the proper combinations to choose for a discussion of the orbitals of the complex of two ethylenes; they are not symmetric or antisymmetric under reflection in plane 2. Again the obvious combinations are $\pi_1+\pi_2$ and $\pi_1-\pi_2$ *(12)*.

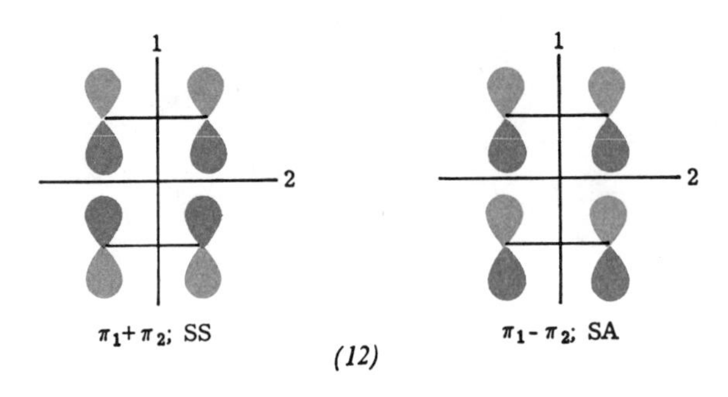

(12)

The first of these is symmetric with respect to reflection in both planes 1 and 2 (abbreviated as S_1S_2 or simply SS); the second is symmetric under reflection in plane 1 and antisymmetric under reflection in plane 2 (S_1A_2 or SA)[9]. Both orbitals, of course, are symmetric with respect to reflection in plane 3, and this trivial information need not be explicitly specified. At large separation between the ethylenes, $\pi_1+\pi_2$ and $\pi_1-\pi_2$ will be degenerate, but at small separations $\pi_1+\pi_2$ will be at low-

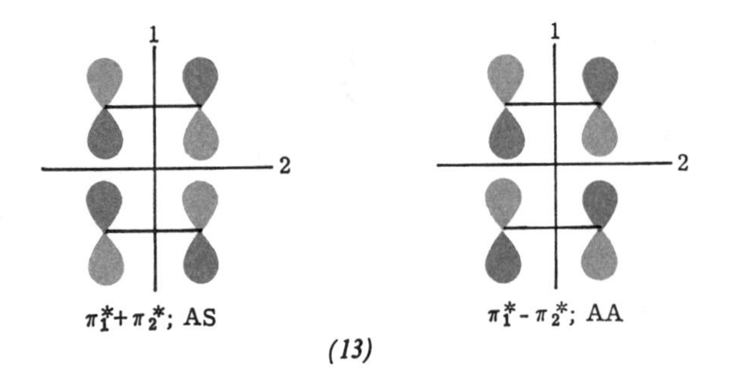

(13)

er energy than $\pi_1-\pi_2$ since the former has fewer nodes. Similar combinations of antibonding molecular orbitals for the complex are shown in *(13)*.

One must next analyze the situation in cyclobutane in an entirely analogous way. Consider the localized σ bonds *(14)*. Again these do not satisfy all the symmetry

[9] We could, of course, label the levels with their proper symmetry designations, appropriate to the D_{2h} symmetry of the approach. We deliberately use the symmetric (S) and antisymmetric (A) labels since the nodal properties of the orbitals are then most clearly discernible.

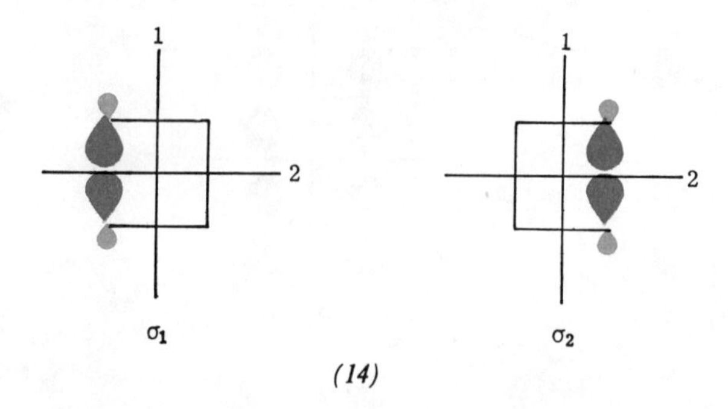

(14)

operations of the cyclobutane molecule, and one must take the two combinations $\sigma_1 \pm \sigma_2$ *(15)*; a similar procedure must be followed for the antibonding σ^* orbitals *(16)*.

We are now equipped to examine the correlation of the orbitals of reactants with those of the product (Figure 7). The direction in which the various levels will move

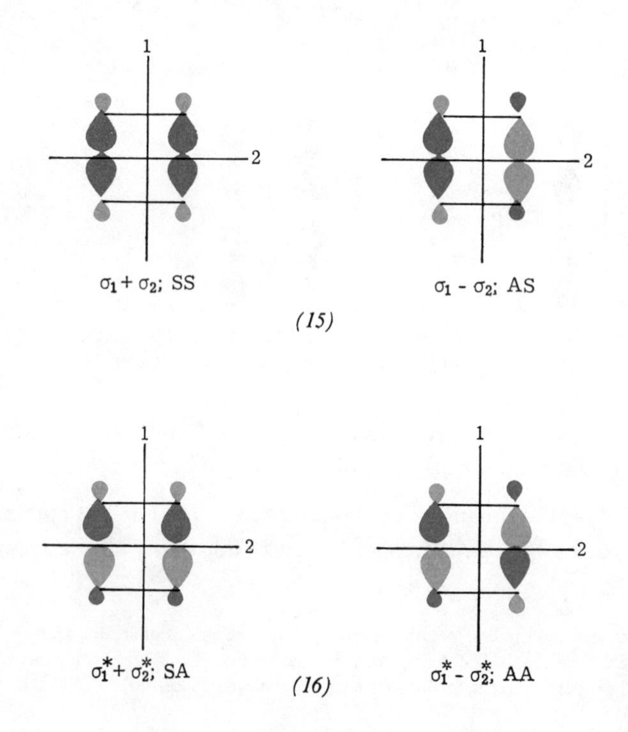

(15)

(16)

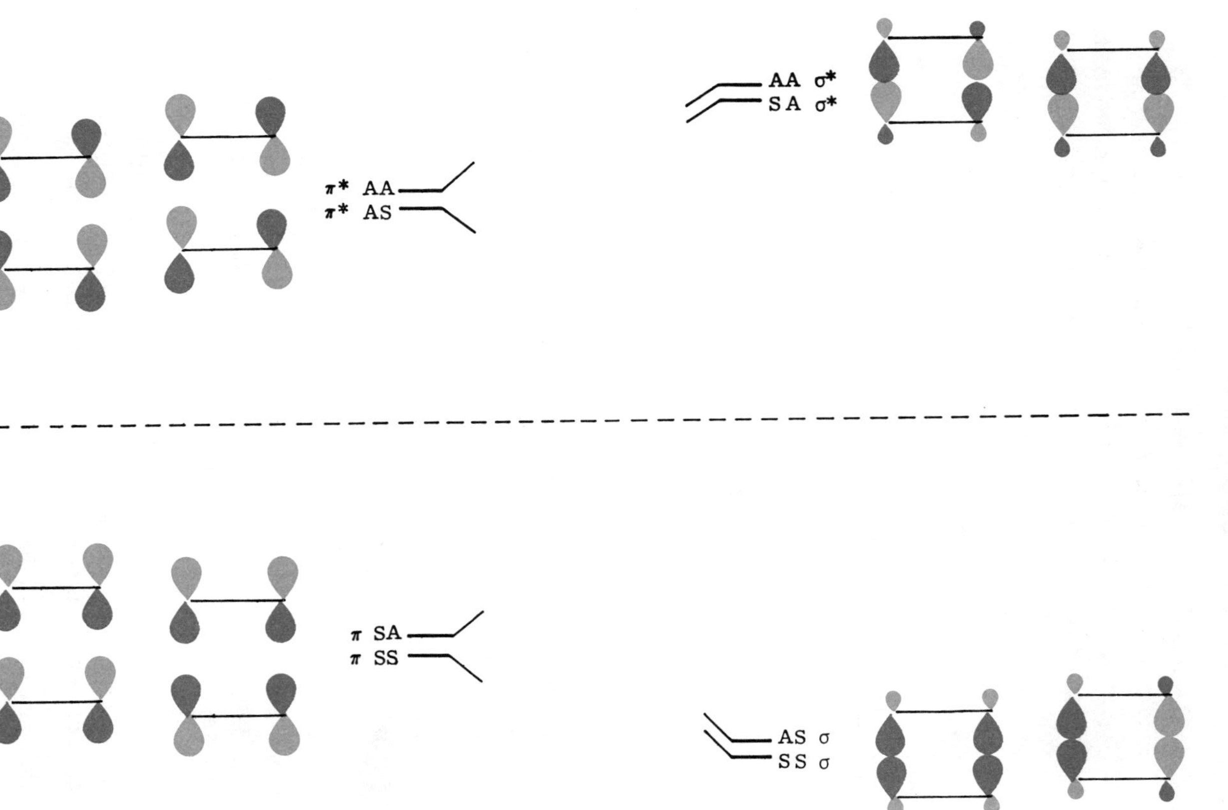

Figure 7. Initial displacements of orbital energy levels in the reactions: 2 ethylenes $\rightleftarrows$ cyclobutane.

may be obtained without detailed calculation, by examining in each case whether any level is bonding or antibonding along the reaction coordinate; in doing this one should keep in mind the definition of bonding or antibonding. Consider the two molecular orbitals for the hydrogen molecule, formed from two 1*s* orbitals

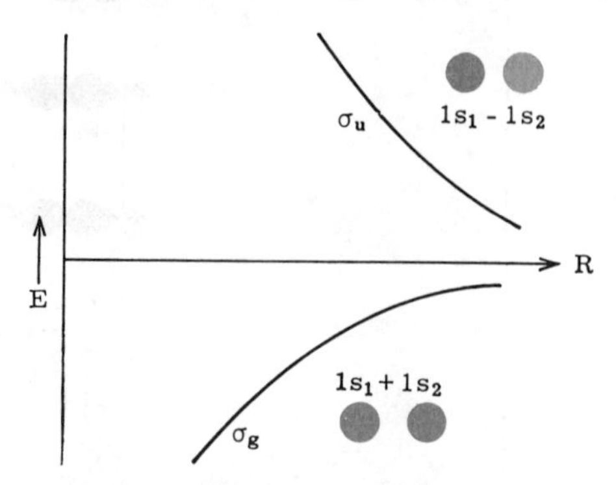

Figure 8. Formation of the bonding (σ_g) and antibonding (σ_u) molecular orbitals of the hydrogen molecule from two atomic 1*s* orbitals.

(Figure 8). σ_g is bonding because electrons occupying it lie in the region between and are shared by the nuclei; σ_u is antibonding and at higher energy because the presence of a node between the nuclei isolates electrons populating the orbital in the regions of the individual terminal nuclei. There is another aspect of bonding or antibonding displayed in this schematic diagram: electrons placed in a bonding orbital bring the nuclei closer together (that is, for σ_g, $\partial E/\partial R>0$), while electrons put into an antibonding orbital push the nuclei apart (that is, for σ_u, $\partial E/\partial R<0$). Returning to the ethylene correlation diagram (Figure 7), we find that the lowest SS level of two ethylenes is bonding in the region of approach of the two molecules to each other and thus will be stabilized by interaction. The SA level has a node, and consequently is antibonding in the region of approach. At large distances the interaction is inconsequential, but as the distance between the reacting molecules diminishes, this orbital is destabilized and moves to higher energy. Similarly, the antibonding π^* AS orbital becomes bonding in the region of approach. It will thus be stabilized as the reaction proceeds, while the π^* AA orbital will be destabilized.

On the cyclobutane side both the σ levels, SS and AS, are bonding in the region where the cyclobutane is being pulled apart. Thus, they resist the motion — that is, they are destabilized along the reaction coordinate. On the other hand, the σ^* levels

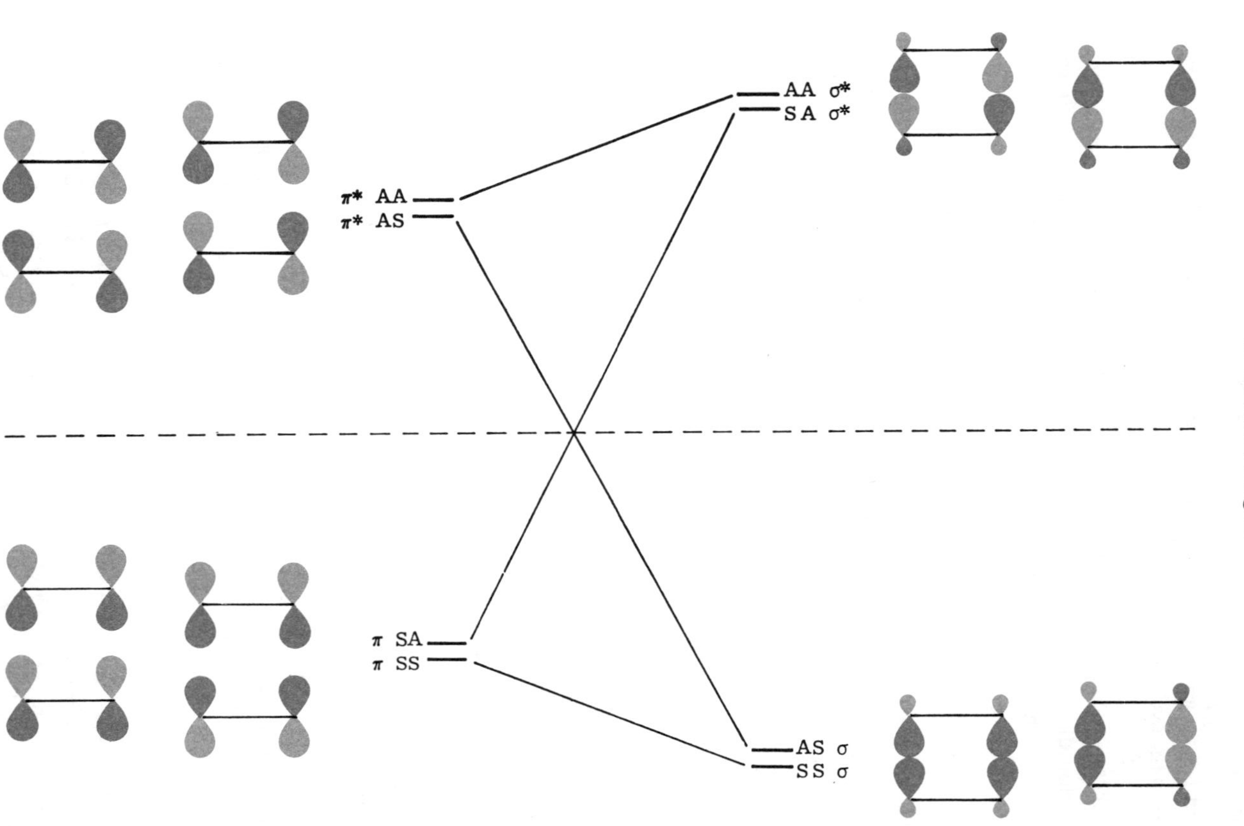

Figure 9. Complete correlation diagram for the formation of cyclobutane from two molecules of ethylene.

SA and AA are antibonding along the reaction coordinate and thus move to lower energy as the cyclobutane is pulled apart.

That these qualitative conclusions are correct may be seen from a completed correlation diagram in which levels of like symmetry are connected (Figure 9). The most obvious and striking feature of this diagram is the correlation of a bonding reactant level with an antibonding product level, and *vice versa.*

We now approach a central tenet of our treatment of concerted reactions. Clearly, *if orbital symmetry is to be conserved*, two ground-state ethylene molecules cannot combine in a concerted reaction to give ground-state cyclobutane — nor can cyclobutane be decomposed in a concerted fashion to two ethylene molecules — through a transition state having the geometry assumed here. To put the matter in other words, there is a very large symmetry-imposed barrier to the reaction under discussion, in either direction. By the same token, there is no such symmetry-imposed barrier to the reaction of one molecule of ethylene with another, one of whose electrons has been promoted, say by photochemical excitation, to the lowest antibonding orbital. For these reasons we designate reactions of the first type *symmetry-forbidden*, and those of the second *symmetry-allowed.*

The matter may be further illuminated by inspection of the corresponding state diagram for the reaction (Figure 10). The ground state electron configuration of two ethylene molecules correlates with a very high-energy doubly excited state of cyclobutane; conversely the ground state of cyclobutane correlates with a doubly excited state of two ethylenes. Electron interaction will prevent the resulting crossing, and force a correlation of ground state with ground state. But in the actual physical situation, the reaction still must pay the price in activation energy for the intended but avoided crossing. An order of magnitude estimate of the symmetry-imposed energy barrier to the concerted face-to-face combination of two ethylene molecules may be made by considering the energy required to raise two bonding electrons in the occupied bonding levels to the non-bonding level — perhaps 5 eV or about 115 kcal/mole.

The lowest excited state of two ethylenes, the configuration $(SS)^2(SA)^1(AS)^1$, correlates directly with the first excited state of cyclobutane. Consequently, there is no symmetry-imposed barrier to this transformation. This represents the course which is followed in many photochemical transformations. However, it should be emphasized that there are ambiguities in excited state reactions which do not exist

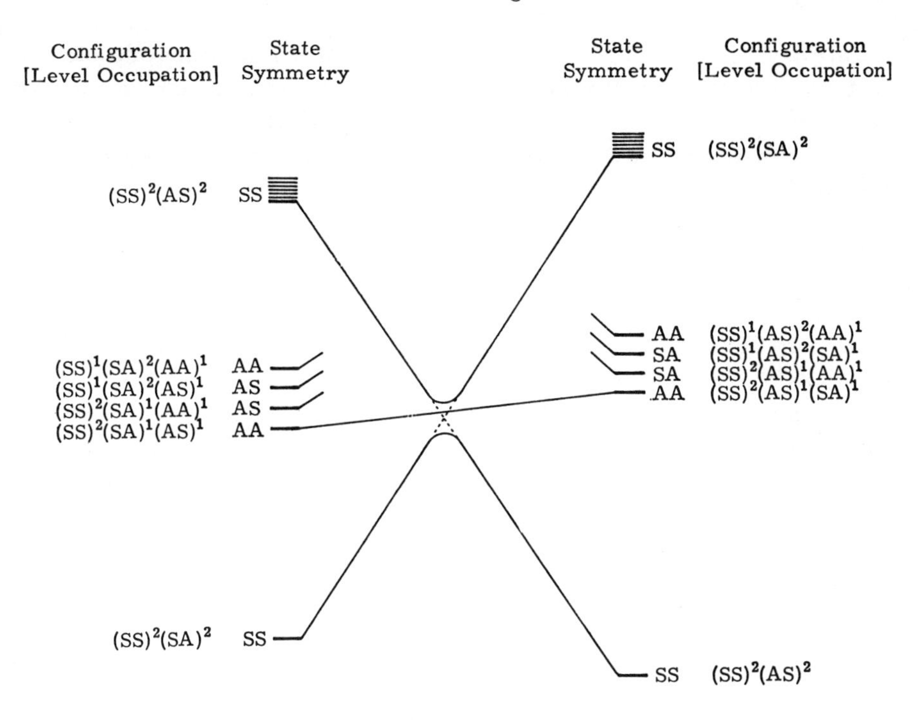

Figure 10. Electronic state diagram for the formation of cyclobutane from two molecules of ethylene.—Note that the symmetry of the states is obtained by multiplying the symmetry labels for each electron, following the rules[7]

$$S\times S\rightarrow S\leftarrow A\times A$$
$$S\times A\rightarrow A\leftarrow A\times S$$

Only singlet excited states are shown, and at this level of sophistication they should be regarded as degenerate in both reactants and product.

in their simpler thermal counterparts. Thus, it may happen that the chemically reactive excited state is not that reached on initial excitation; in particular, singlet-triplet splittings for different excited states vary so widely that the symmetries of the lowest singlet and lowest triplet states may differ. Further, radiationless decay may be so efficient that the chemical changes subsequent to irradiation may be those of a vibrationally excited ground state. Finally, the formation of a transition state for a given concerted reaction may be competitive with relaxation of the excited state component to an equilibrium geometry which renders the reaction geometrically impossible. It should be emphasized that none of these punctilios in any way vitiates the consequences of orbital symmetry control. The principle of conservation of orbital symmetry remains applicable, *provided that the chemically reactive excited state is identified;* furthermore, the fact that the product state which correlates directly with the reactant state may lie higher in energy than the latter constitutes no special problem in orbital symmetry terms — though admittedly there is still much to be learned about the detailed physical nature of the processes accompanying the necessary energy cascade from electronically excited to ground states in such instances.

Many correlation diagrams differ sharply from that for the ethylene + ethylene reaction. Consider for example the prototype Diels-Alder reaction — the [4+2] cycloaddition of butadiene to ethylene. The most reasonable symmetric approach is characterized by a single plane of symmetry bisecting the two components (Figure 11). The essential levels involved in the reaction are now six in number, and they

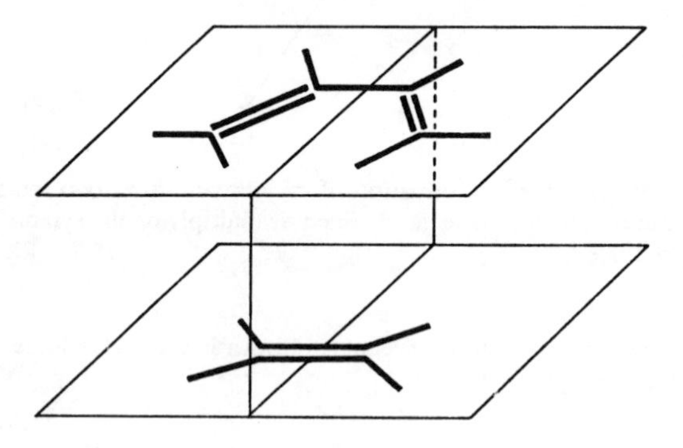

Figure 11. Symmetric approach of butadiene and ethylene in the Diels-Alder reaction.

are illustrated in the correlation diagram shown in Figure 12. The form of the four butadiene and two ethylene orbitals is self-evident. We have placed the ethylene π level between the two bonding diene orbitals, but the ordering is not consequential. On the product cyclohexene side note that one must construct delocalized σ bond combinations just as was done for cyclobutane.

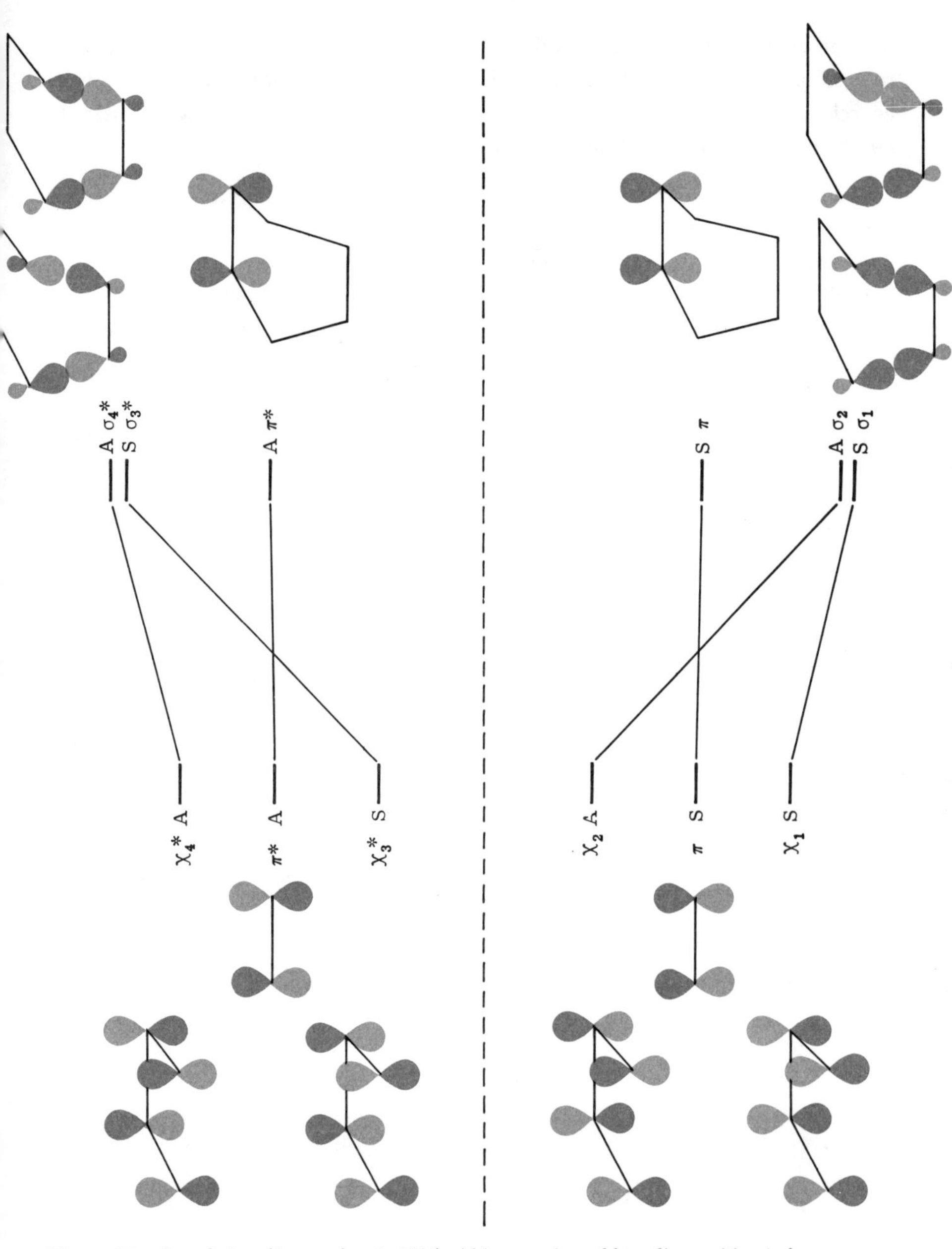

Figure 12. Correlation diagram for the Diels-Alder reaction of butadiene with ethylene.

The difference between this correlation diagram and that for the combination of two molecules of ethylene is striking. In this case every bonding level of reactants correlates with a bonding product level; there is no correlation which crosses the large energy gap between bonding and antibonding levels.

As before one may construct a state diagram (Figure 13). The ground state levels correlate directly, and the diagram implies that there is no activation energy at all for this thermal symmetry-allowed process. And indeed, *there is no symmetry-imposed barrier*, but of course there is an activation energy — experimentally found to approximate 20 kcal/mole — which arises from factors not simply related to orbital

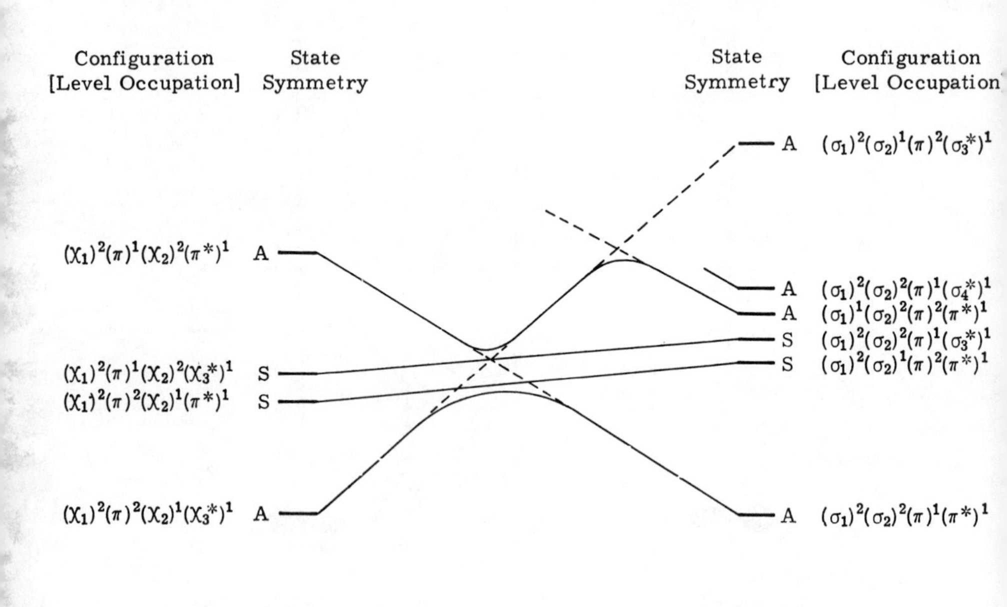

Figure 13. Electronic state diagram for the Diels-Alder reaction of butadiene with ethylene.

symmetry conservation; among them are energy changes accompanying rehybridization in the levels we have not included, and bond length extensions and contractions, as well as angle distortions.

The first excited state of the diene-ethylene complex does not correlate with the first excited state, $\pi \rightarrow \pi^*$, of cyclohexene. Consequently, in this case a symmetry-imposed barrier arises in the excited state process. These circumstances are perhaps most simply apprehended when it is realized that the first excited state is formed by promotion of an electron from an orbital decreasing in energy along the reaction coordinate to one increasing in energy.

The physical correlations of some levels may seem non-intuitive, but in each case they are realistic, and can be understood if followed through carefully in detail. For instance, it may seem strange that an ethylene π level becomes a π level of the cyclohexene. To understand this relationship one must allow the ethylene π level to interact with the two other symmetric levels — the lowest occupied and lowest unoccupied diene levels. An important general rule of quantum mechanics is relevant here[10]: if two levels of unlike energy interact, that of the lower energy will mix into itself some of the higher-energy wave function in a bonding way, but the higher level will mix into itself some of the lower one in an antibonding way. If more than two levels interact, their mixing can be analyzed as a superposition of such pair-wise interactions. The application of this rule is illustrated here for the formation of a C-H σ bond from a carbon sp^3 hybrid and a hydrogen $1s$ orbital *(17)*. In the case of butadiene + ethylene we find that the olefin π level mixes into

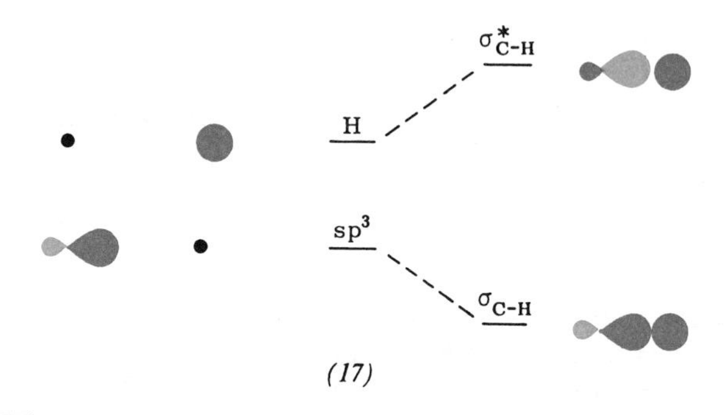

(17)

[10] The rule follows directly from perturbation theory and the correlation of higher energy with increasing number of nodes in a wave function.

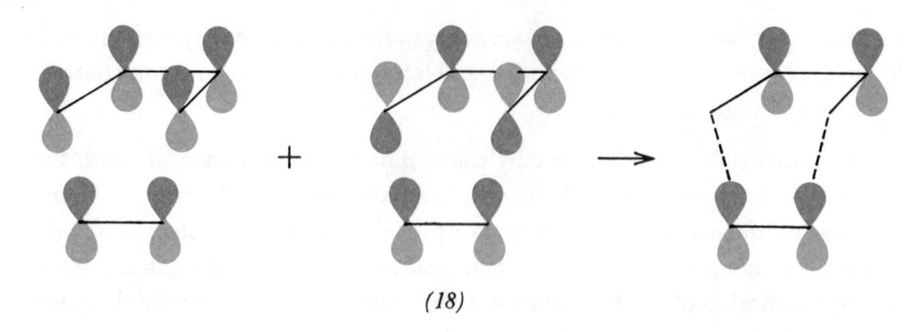

(18)

itself χ_1 in an antibonding way and χ_3 in a bonding way *(18)*. The diene contributions cancel at C-1 and C-4 of the butadiene, but reinforce at C-2 and C-3. Thus, in the transition state this orbital is essentially half in the one and half in the other of the reacting moieties.

If one considers further cases of the general cycloaddition reaction of an m π-electron system with an n π-electron system to form two new σ bonds, while maintaining a plane of symmetry, it becomes evident that there are only two types of correlation diagrams:

a) those similar to that of the Diels-Alder reaction, with no correlation of bonding and antibonding levels, and characterized as symmetry-allowed for ground states and symmetry-forbidden for excited states;

b) those similar to that for the ethylene + ethylene combination, which display bonding-antibonding correlations, and are consequently symmetry-forbidden for ground states and symmetry-allowed for excited states.

To derive a general rule, one may enumerate, say, bonding symmetric levels in reactants and products. If there are, for example, m π orbitals in a reactant, there will be $m/4$ symmetric bonding π orbitals if $m/2$ is even, or $(m+2)/4$ if $m/2$ is odd. That part of the product derived from this component will contain $(m-2)$ π orbitals, of which $m/4$ are symmetric bonding levels if $m/2$ is even, or $(m-2)/4$ if $m/2$ is odd. There are three possible cases (q_1 and q_2 are integers $= 0, 1, 2 \ldots$):

Case	m	n	Total Symmetric Bonding π Levels Before	After
1	$4q_1$	$4q_2$	q_1+q_2	q_1+q_2
2	$4q_1+2$	$4q_2$	q_1+q_2+1	q_1+q_2
3	$4q_1+2$	$4q_2+2$	q_1+q_2+2	q_1+q_2

Of the new bonding σ levels in the product, one is always symmetric; thus, for a thermal symmetry-allowed reaction, the total number of occupied symmetric π bonding levels in the reactants must exceed by one the number in the product; Case 2 satisfies this condition: for it

$$m+n = 4q_1+4q_2+2 = 4q+2.$$

Cases 1 and 3 necessarily lead to bonding-antibonding correlations; consequently, thermal reactions in which

$$m+n = 4q_1+4q_2 \text{ or } 4q_1+4q_2+4 = 4q$$

are symmetry-forbidden. In each case, of course, these rules are precisely reversed for reactions involving excited states. Further, it must be emphasized that these deductions apply only to cases in which the geometry of approach of the reacting molecules is that specified above; we shall demonstrate in the sequel that in some cases the same, and in others different, rules apply when the geometric relationships within the transition state are varied.

3.1. General Comments on the Construction of Correlation Diagrams

A crucial phase in the construction of a correlation diagram is the identification of pertinent σ and π levels and their delocalization to the full extent required by the transition state symmetry. The simplest procedure may be formulated as follows:

a) Identify all orbitals involved as σ, π, or n (nonbonded pair). Remember that for each σ and π level there will be corresponding σ^* and π^* levels; this is not the case for n orbitals. Thus, for the expulsion of carbon monoxide from cyclopentenone *(19)* the relevant orbitals are: in the cyclopentenone the C=C π and π^* levels, two bonding C-C(O) σ levels and the corresponding σ^* orbitals, and a lone pair on oxy-

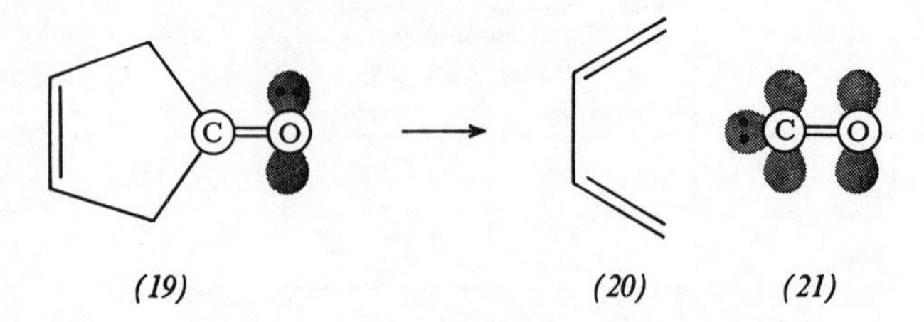

(19) *(20)* *(21)*

gen; in the products *(20)* and *(21)* the four diene π levels, two of which are bonding, a new CO bond and its π^* counterpart, and a new nonbonding pair level on the carbon atom of the carbon monoxide. Left out of consideration as non-essential are another oxygen lone pair and the remaining C=O π bond. Note that there exists an automatic check in that the number of levels of each symmetry type on the right-hand side of a correlation diagram must equal that on the left.

(22)

b) If a polyene system is present, the proper molecular orbitals of a polyene should be used. Thus, in the cycloaddition of a heptafulvalene *(22)*, one should consider the orbitals of a fourteen-membered polyene.

(23) *(24)*

For the cyclization of benzene *(23)* to prismane *(24)*, the proper orbitals on the left are those of benzene [*cf.* Section 6.4].

(25) *(26)*

(27) *(28)*

However, if only fragments of a polyene system participate in a reaction, the molecular orbitals of the component fragments must be used. Thus, the conversion of a hexatriene *(25)* to a bicyclo[3.1.0]hexene *(26)*, or to a bicyclo[2.2.0]hexene *(28)* must be treated as [4+2] and [2+2] cycloadditions respectively.

c) Mix all bonding σ orbitals which are not symmetric or antisymmetric with respect to every molecular symmetry element until they become so. For the vast majority of cases this process is one of identifying the symmetry-related orbitals, and forming their sums and differences. The procedure should then be repeated for σ^* levels. As one example, consider the newly-formed σ bonds of a cyclohexene—the product of a [4+2] cycloaddition. The symmetry-related σ orbitals are σ_1 and σ_2; they must be transformed as indicated in *(29)*, to $\sigma_1+\sigma_2$ and $\sigma_1-\sigma_2$.

Again, consider the σ^* orbitals of cyclopentenone *(30)*.

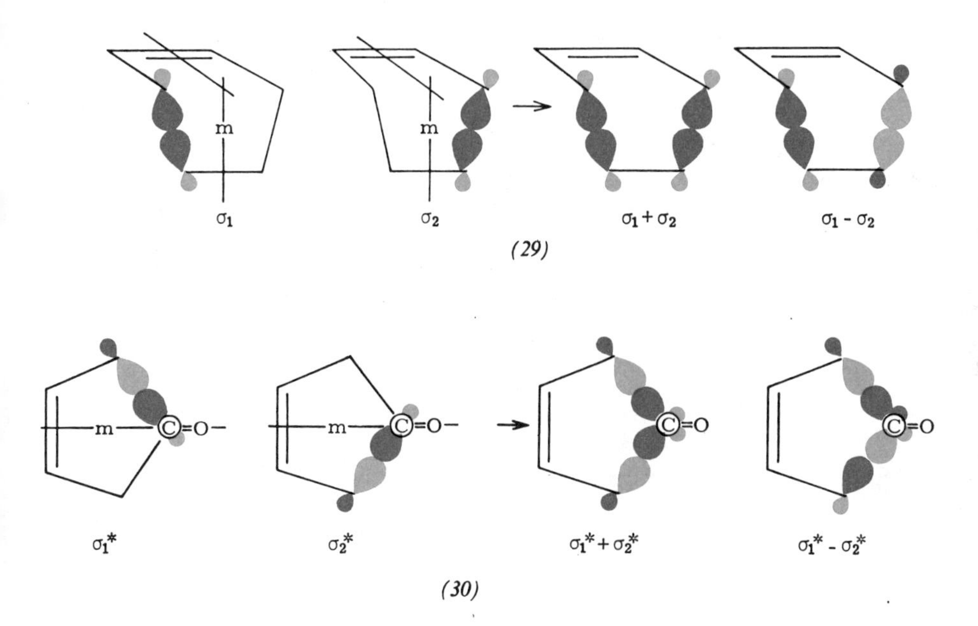

(29)

(30)

A similar mixing is required for nonconjugated π levels. Thus, in considering a [2+2] cycloaddition within a 1,5-hexadiene, leading to a bicyclohexane, the preparation of π orbitals is performed as shown in *(31)*.

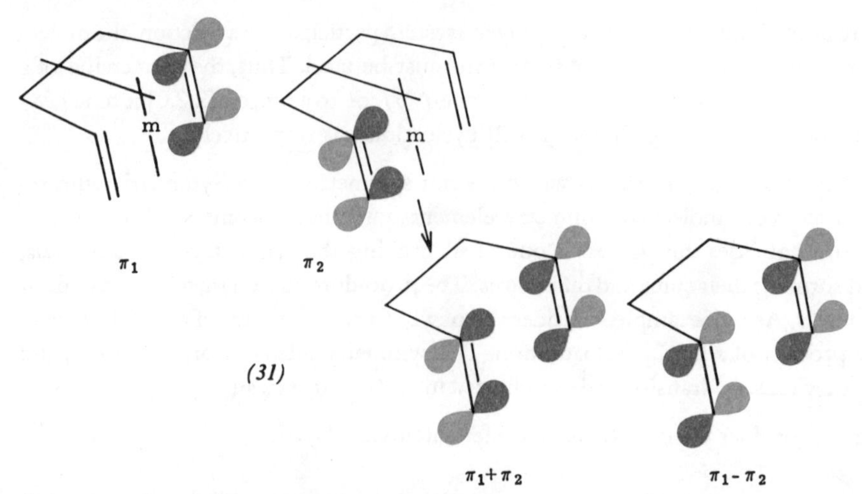

(31)

d) A convenient, but not necessary, further step is to mix with each other all orbitals of a given symmetry. For example, consider the σ bonds of a [2+2+2] cycloreversion of a cyclohexane to three ethylenes. Assume a boat conformation of the

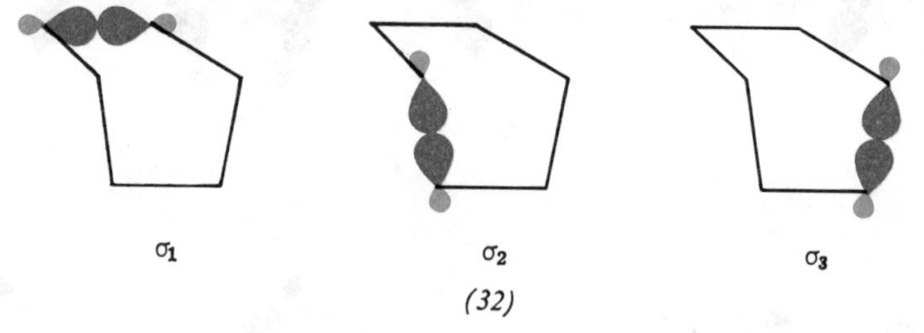

(32)

cyclohexane. The three localized σ orbitals are σ_1, σ_2, and σ_3 *(32)*. The only symmetry element in the transition state is a plane bisecting σ_1. The orbital σ_1 is symmetric under reflection in that plane, but σ_2 and σ_3 are not. Thus, in accordance

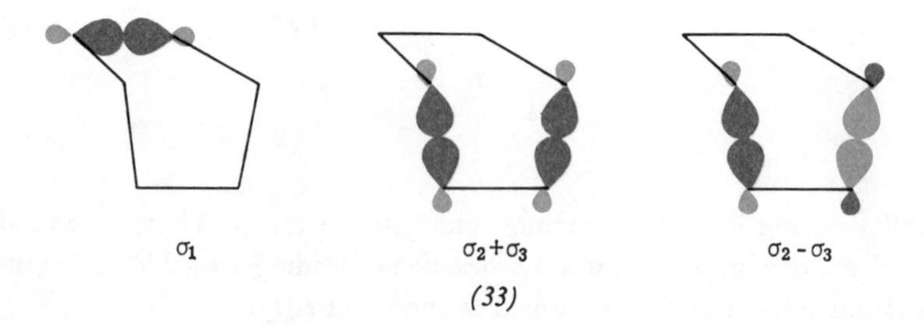

(33)

with rule c) above, we should form the delocalized combinations $\sigma_2+\sigma_3$ and $\sigma_2-\sigma_3$. This yields a new set of symmetry-adapted orbitals *(33)*.

Now σ_1 and $\sigma_2+\sigma_3$ are symmetric with respect to the symmetry plane, and it is convenient to delocalize further by taking their sum and difference *(34)*.

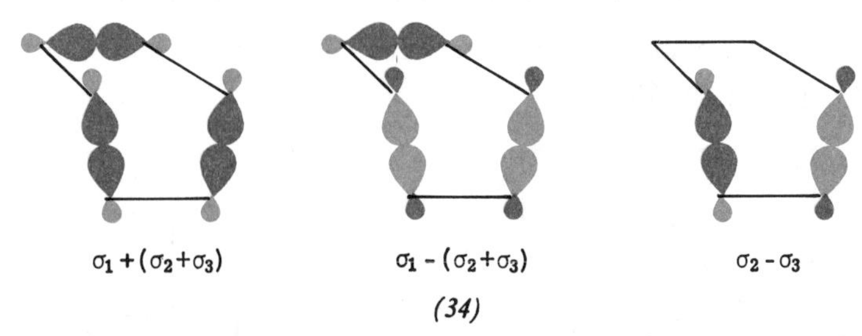

(34)

More extensive delocalization (a secondary effect — but one which can be of chemical significance) would follow from mixing of σ_1^* and $\sigma_2^*-\sigma_3^*$ into $\sigma_2-\sigma_3$, and $\sigma_2^*+\sigma_3^*$ into the symmetric combinations.

3.2. Precautions in the Construction of Correlation Diagrams

There are several pitfalls in the construction and application of correlation diagrams. To avoid these the following precautions must be observed:

a) Each basic process must be isolated and analyzed separately. Otherwise the superposition of two forbidden but independent processes may lead one to the erroneous conclusion that the combined process is symmetry-allowed. Two wrongs do not make a right.

b) The symmetry elements chosen for analysis must bisect bonds made or broken in the process. Here there are two corollaries:

(i) a symmetry element of no use in analyzing a reaction is one with respect to which the orbitals considered are either all symmetric or all antisymmetric — obviously an analysis based on such elements only (such as plane 3 in the approach of two ethylenes [*cf.* Figure 5]) would lead to the conclusion that every reaction is symmetry-allowed;

(ii) if the only symmetry element is one which does not bisect any bonds made or broken, then the correlation diagram constructed on the basis of this element can only lead to the conclusion — often false — that a reaction is symmetry-allowed.

c) Each case must be reduced to its highest inherent symmetry. Thus, if there are heteroatoms in a polyene component, they are to be replaced by their isoelectronic carbon groupings. If there are substituents with trivial electronic demands they should be replaced by hydrogens. Heteroatoms do offer the possibility of new reactions by the inclusion of nonbonding pairs or by the availability of low-lying unoccupied orbitals. These types of interaction should be carefully analyzed.

To consider an example of a), take the formation of cubane from cyclooctatetraene in one step (Figure 14).

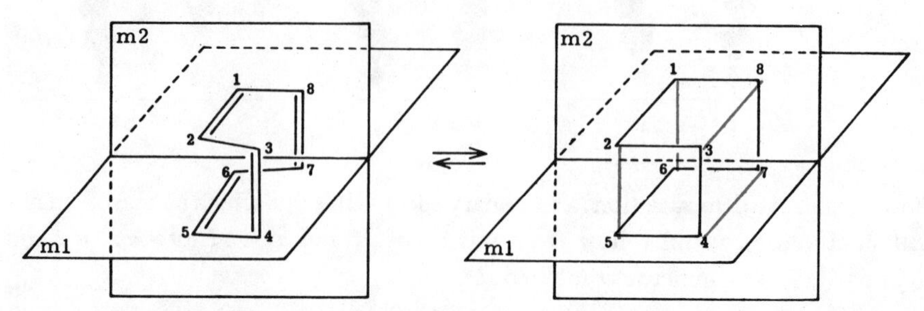

Figure 14. Formation of cubane from cyclooctatetraene in a concerted reaction.

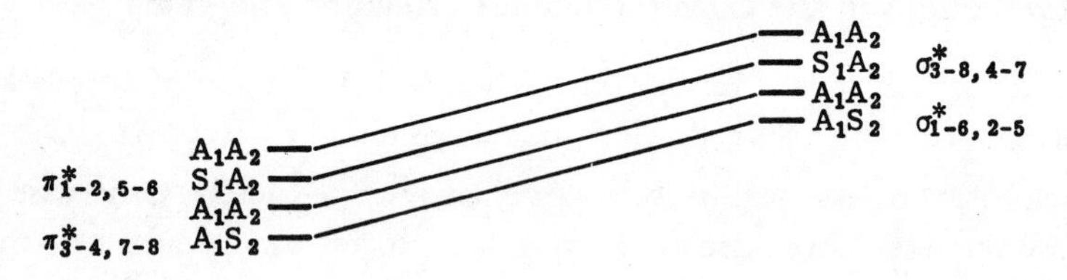

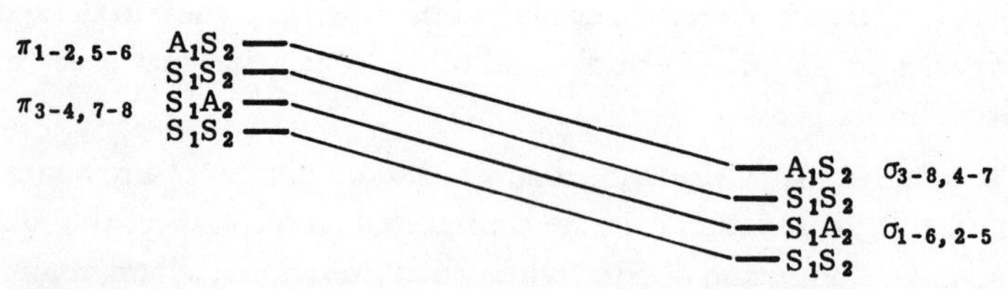

Figure 15. Incorrect correlation diagram for the formation of cubane from cyclooctatetraene.

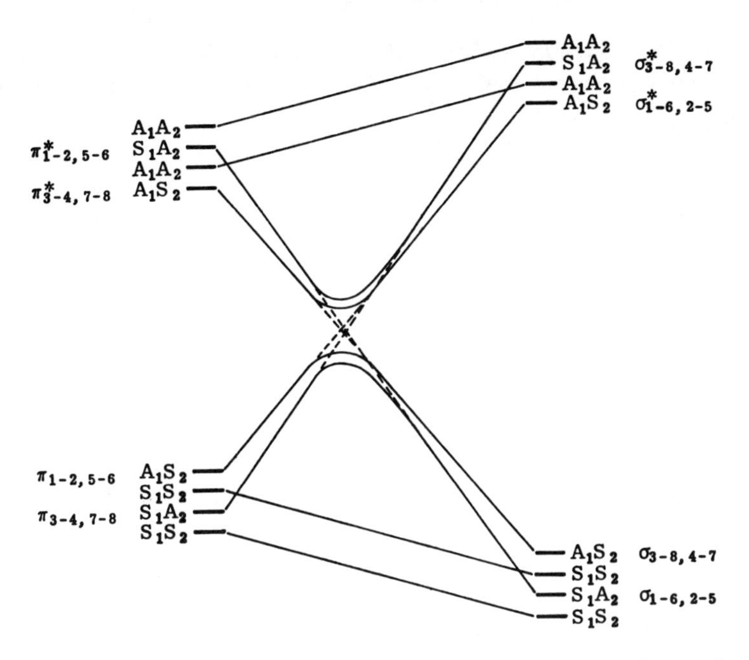

Figure 16. Correct correlation diagram for the formation of cubane from cyclooctatetraene. — Within any of the four groups of four closely spaced levels, those with both symmetric and antisymmetric labels, regardless of subscripts, are in fact degenerate, though shown here for clarity as differing slightly in energy.

Construction of a correlation diagram (Figure 15) without regard to precaution a) would lead to the conclusion that this superposition of two [2+2] cycloadditions should be a symmetry-allowed thermal process.

In fact the diagram should really appear as in Figure 16. The intended correlation of the S_1A_2 orbital formed from $\pi_{3\text{-}4,7\text{-}8}$ is to an antibonding S_1A_2 $\sigma^*_{3\text{-}8,4\text{-}7}$ orbital. A small perturbation links the 3-4,7-8 cycloaddition with the 1-2,5-6 process. The latter by an accident of symmetry is forming an S_1A_2 orbital of the cubane. The result is that the intended bonding-antibonding correlation is accidentally avoided. But the reaction is none the less symmetry-forbidden. In fact, whether a reaction is symmetry-allowed or forbidden is determined by the height of the electronic hill that reactant or product orbitals must climb in reaching the transition state; and the presence or absence of a hill is a function of the *intended* correlation, or the initial slope of the levels. From this point of view, we see that in the case at hand the 3-4,7-8 combination is in no wise facilitated by the concurrent 1-2,5-6 process, and *vice versa.*

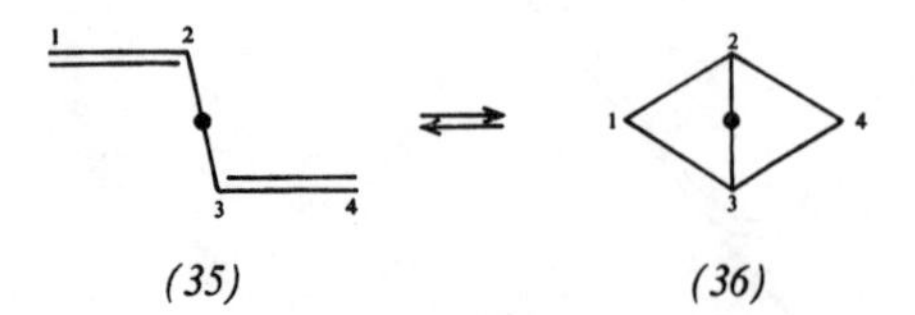

(35) *(36)*

An example of type b) is provided by consideration of the conversion of butadiene *(35)* to bicyclobutane *(36)*. Here the only molecular symmetry element is a two-fold axis passing through a single bond which is not made or broken during the reaction. We will discuss this reaction in some detail in the sequel.

(37) *(38)*

Another instance is the case of two propylenes trading hydrogens [*(37)*→*(38)*], where only a center of symmetry (passing through no bonds) could be present.

(39) *(40)*

Still another example is the decomposition of pentalene to diacetylene and two acetylenes [*(39)*→*(40)*], where no element of symmetry bisects bonds broken or made.

The rationale for point c) will become apparent on comparison of the correlation diagram for the face-to-face addition of ethylene to ethylene with that for ethylene

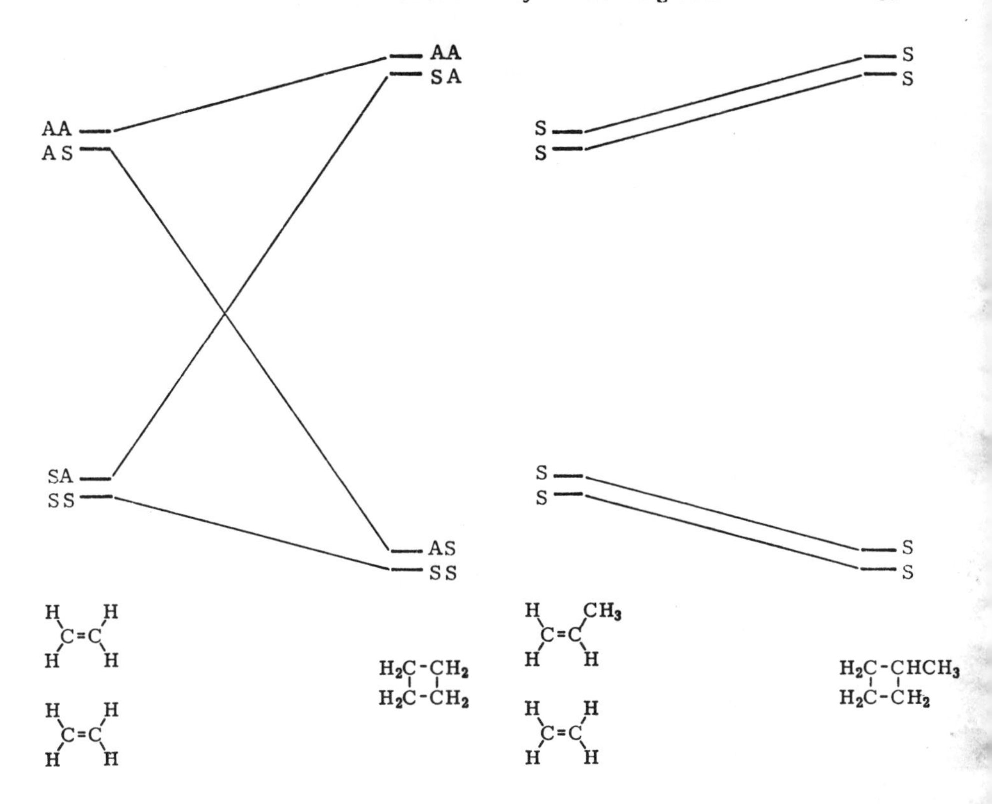

Figure 17. Left: Correlation diagram for the cycloaddition of two ethylene molecules. – Right: Incorrect correlation diagram for the cycloaddition of ethylene to propylene.

to propylene. For the latter case there is, strictly speaking, no symmetry in the transition state. Thus, all levels are trivially symmetric and one might be tempted to draw the correlation diagram as shown at right in Figure 17.

The conclusion would be that methyl substitution had made the thermal [2+2] cycloaddition symmetry-allowed. This is incorrect. The proper level correlation diagram is shown in Figure 18.

Methyl substitution has in an absolute sense reduced all levels to the same symmetry. The level crossing has become impossible. But the perturbation is a very small one and so the crossing is barely avoided. A bonding level of reactants has still

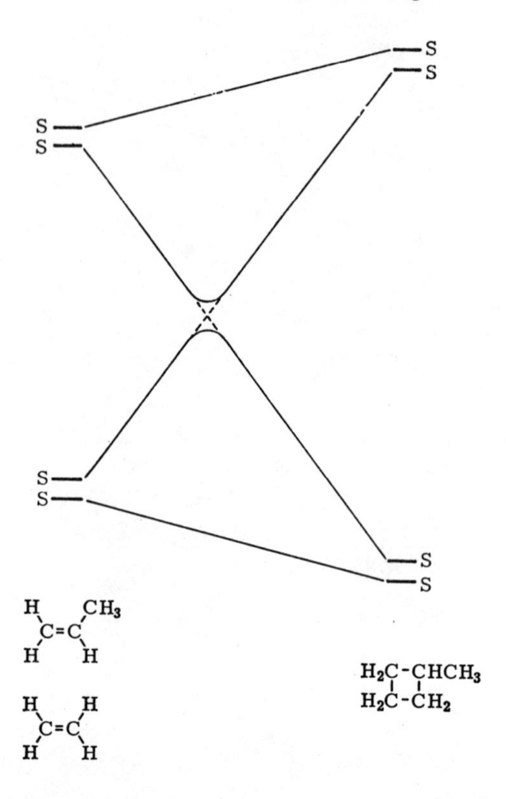

Figure 18. Correct correlation diagram for the cycloaddition of ethylene to propylene.

moved to high energy in the transition state. The corresponding state diagram differs trivially from that for ethylene + ethylene. The reaction remains symmetry-forbidden[11].

---

[11] A similar situation arises in other areas of chemistry. The n→π* transition in formaldehyde is electric-dipole forbidden. The source of the small intensity observed is still disputed. In the cases of acetaldehyde, or of unsymmetrically substituted ketones, the symmetry element which made the formaldehyde transition electric-dipole forbidden is removed. The transition becomes allowed. Does it therefore jump to a high intensity? Not at all; the intensity remains practically unchanged. This is because the *essential* symmetry, that of the local environment of the carbonyl group, is unchanged.

4. The Conservation of Orbital Symmetry

It is clear that the absence or presence of molecular symmetry in an absolute sense cannot be the ultimate source of the allowedness or forbiddenness of a reaction. Symmetry is discontinuous. It is either on or off, here or not here. Chemistry is obviously not like that. A slight perturbation, say substitution by a methyl group, may destroy total symmetry, but cannot be expected to change dramatically the mechanism of a reaction. The essential and decisive factor in making a reaction forbidden is that in the transition state there is at least one level that is no longer bonding, but at considerably higher energy. We have used symmetry as a crutch to aid us in denominating those high energy levels without doing the least bit of calculation. If symmetry is lacking, either as a result of trivial substitution, or more basically from the asymmetry of the components (*e.g.*, in the "ene" reaction), a reaction may still be analyzed by writing down the orbitals involved, allowing them to mix according to well-defined quantum mechanical principles and following the interacting orbitals through the reaction. High energy levels in the transition state may arise *pari passu* from real crossings or from intended but avoided ones. Such high energy levels will not be present if every bonding orbital of the product (reactants) is derivable from one bonding orbital of the reactants (product). For if a bonding orbital of the product is not derivable from any bonding orbital of reactants, it must be related to an antibonding reactant orbital. Whether or not it actually correlates to that antibonding orbital is dependent on the presence or absence of total symmetry; but even if the intended correlation is foiled, the level soars to high energy in the transition state.

In short, the most general, and at the same time physically most realistic view of orbital symmetry control of chemical reactions is obtained by specifying the relevant molecular orbitals of reactants, and observing their corresponding form in products, as reaction occurs *with conservation of orbital symmetry*. To illustrate our method at work we will first discuss some instances where correlation diagrams can be drawn, one of these in great detail, and then proceed to less symmetrical cases where no help may be expected from molecular symmetry.

5. Theory of Electrocyclic Reactions

These intramolecular cycloadditions provided the stimulus for our study of molecular orbital symmetry and concerted reactions. We define as *electrocyclic* reactions the formation of a single bond between the termini of a linear system containing k π electrons, and the converse process *(41)*.

k ⇄ k-2

(41)

In such changes fixed geometrical isomerism imposed upon the open-chain system is related to rigid tetrahedral isomerism in the cyclic array. *A priori*, this relationship might be *disrotatory* or *conrotatory* *(42)*. In the former case the transition state is characterized by a plane of symmetry while in the latter a two-fold axis of symmetry is preserved.

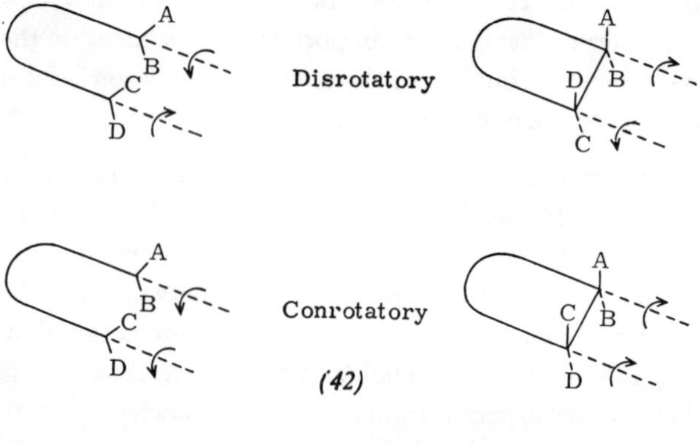

Disrotatory

Conrotatory

(42)

Consider the essential molecular orbitals in the conversion of cyclobutene to butadiene. These are the four π orbitals of the butadiene χ_1, χ_2, χ_3, χ_4, the π and π^* levels of the cyclobutene double bond, and the σ and σ^* orbitals of the single bond to be broken *(43)*.

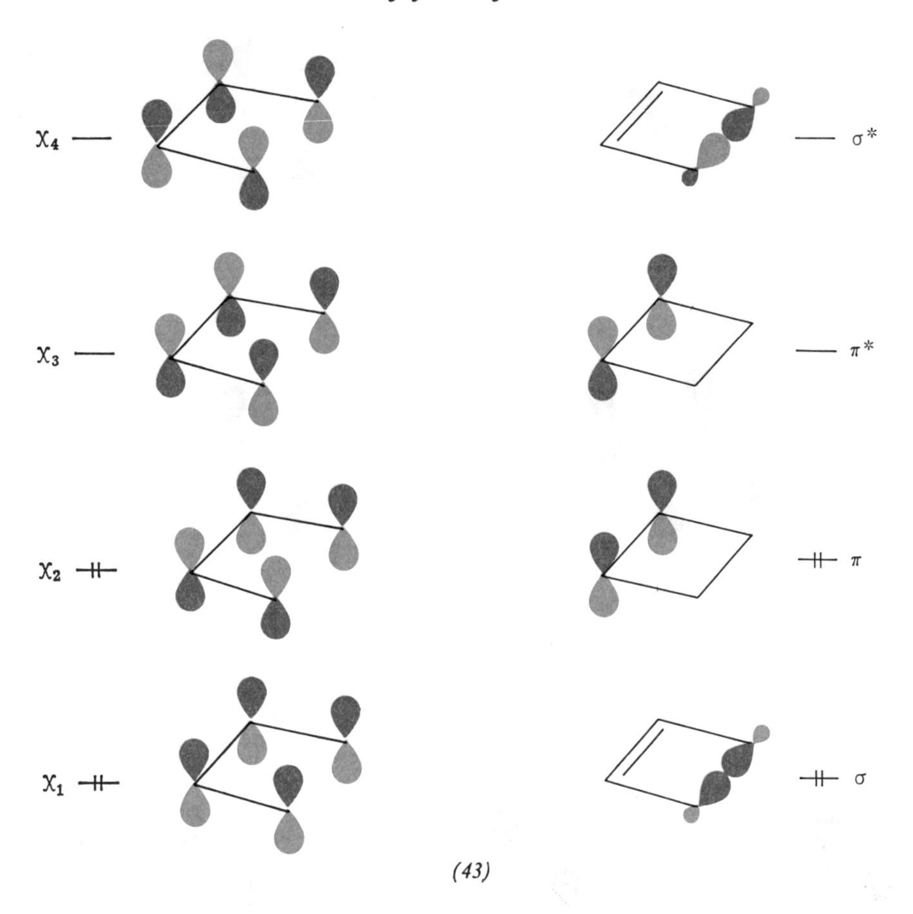

(43)

Consider carrying out a conrotatory motion to completion on σ [step 1 in *(44)*] and follow through with a rehybridization (step 2). At this stage the orbital looks like a fragment of χ_2 (or χ_4) of butadiene and all that is needed is a growing-in (step 3) of orbitals at C-2 and C-3.

It should be kept in mind that in reality the steps 1, 2, and 3 will all be simultaneously proceeding along the reaction coordinate, and that the above factorization is only an aid to visualization. The growing-in of step 3 may seem like magic to those unfamiliar with molecular orbital manipulations. It is in fact a universal phe-

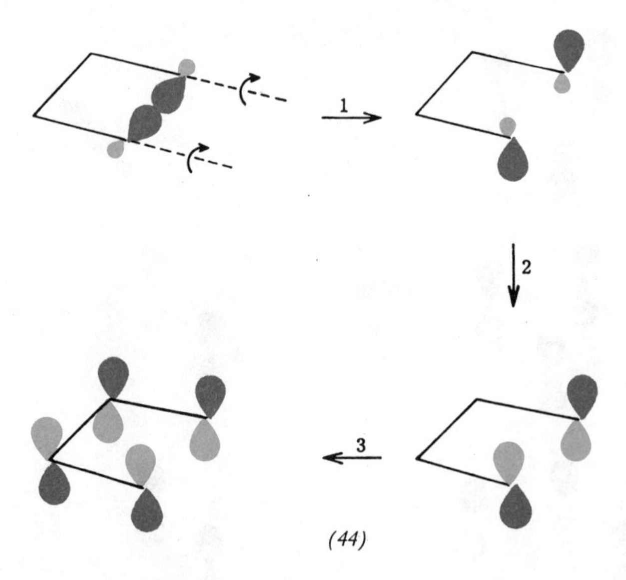

(44)

nomenon, the detailed result in this case of mixing of π^* with σ as the reaction proceeds. We have now followed σ through the reaction and correlated it with another bonding orbital, χ_2. Similarly we follow π through *(45)*. Note here that the

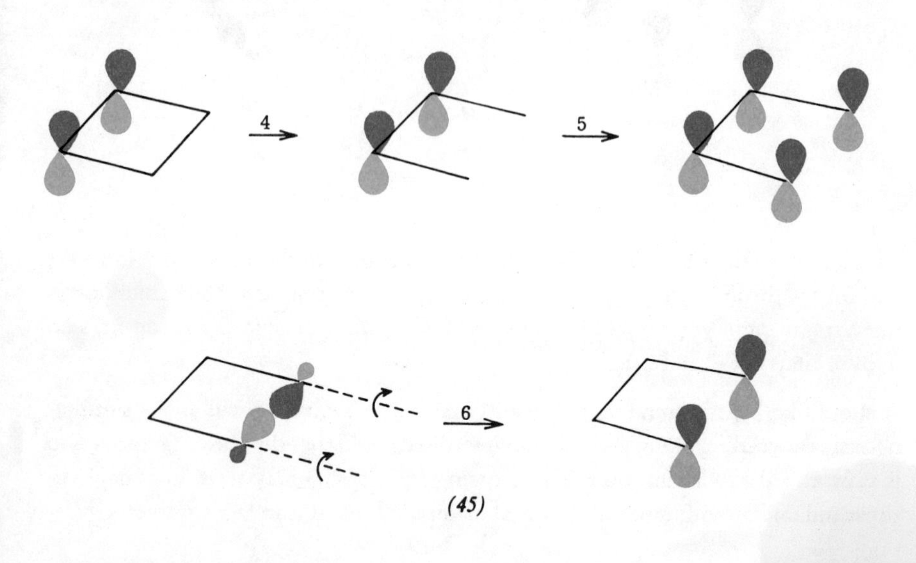

(45)

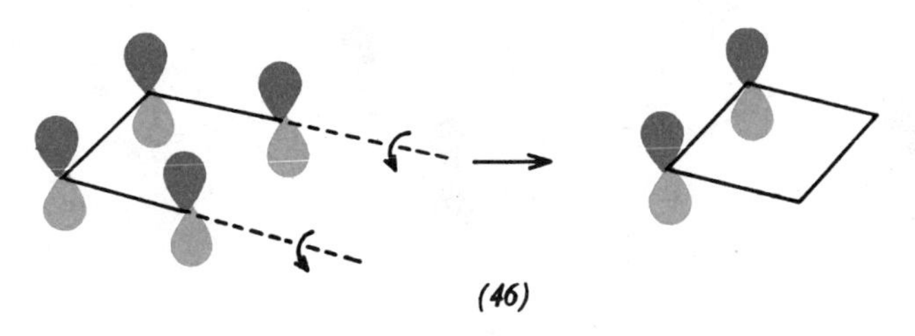

(46)

growing-in step 5 is really a mixing of π with σ^* (*cf.* step 6). π thus correlates with χ_1.

The correlations could of course have been obtained starting from butadiene. χ_1 by a conrotatory motion becomes π *(46)* and χ_2 is transformed into σ *(47)*.

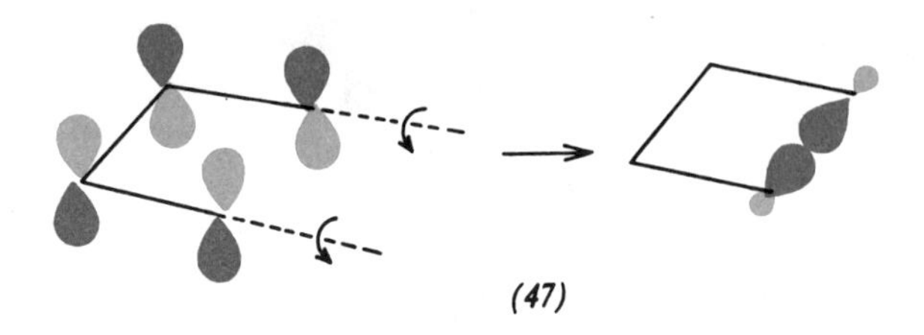

(47)

In this analysis there appears a fading-away, in which extra nodes and contributions disappear. This is the precise reverse of the growing-in noted above and is again a result of mixing with higher orbitals of the proper symmetry. Very similar arguments lead to a correlation of σ^* and χ_3, and of π^* and $-\chi_4$ ($\equiv \chi_4$!). We have thus achieved a correlation of bonding levels of the reactant with bonding levels of the product, with conservation of orbital symmetry. The thermal reaction should be a facile one.

By contrast, consider now a disrotatory opening. The correlations are indicated in *(48)*. Both σ and π must correlate with χ_1 or χ_3 ($\equiv -\chi_3$!). But since only one can correlate with χ_1, the other must go up to χ_3, which is antibonding. Conservation of orbital symmetry requires in this case a high-lying transition state and the thermal reaction is symmetry-forbidden. Again the problem could have been approached from the other side *(49)*.

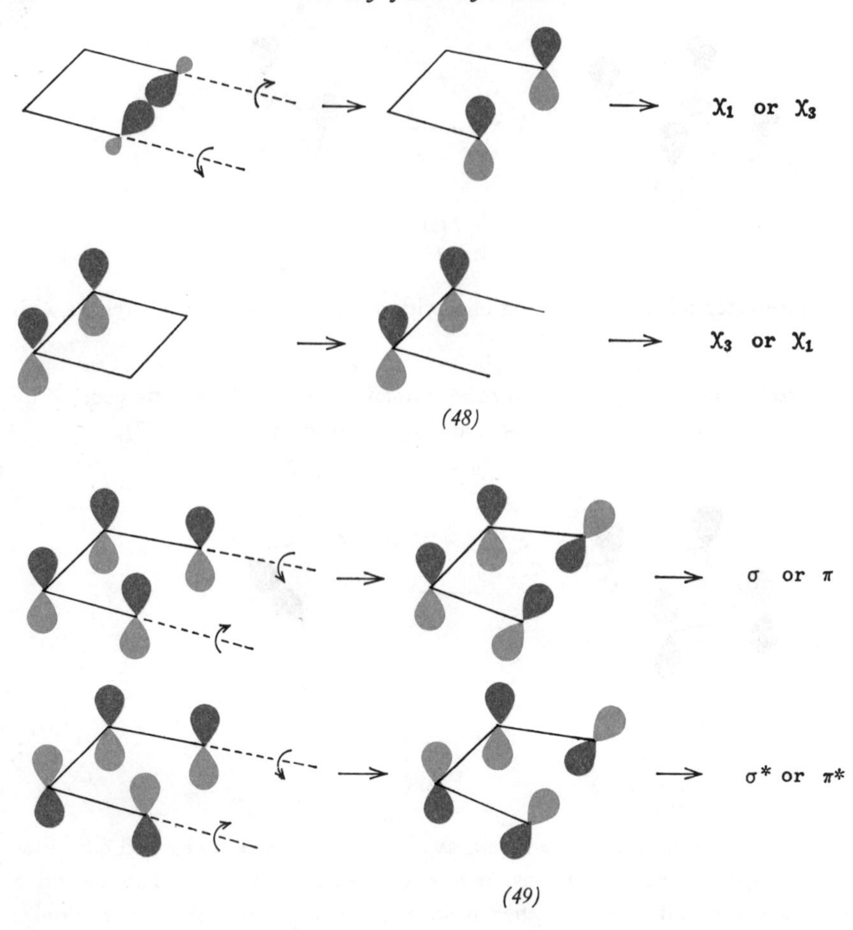

(48)

(49)

It is obvious that χ_2 is the troublesome orbital — it cannot transform into *any* bonding orbital of the cyclobutene, with conservation of orbital symmetry, in a disrotatory process.

What we have somewhat laboriously described in words is equivalent to the construction of two level correlation diagrams (Figure 19). It is clear that in the conrotatory process a two-fold rotation axis is maintained at all times whereas in the disrotatory motion an invariant plane of symmetry is present. These diagrams are clearly analogous to those discussed in Section 3 (above) for combination reactions of π electron systems. Obviously the diagram for the conrotatory process is char-

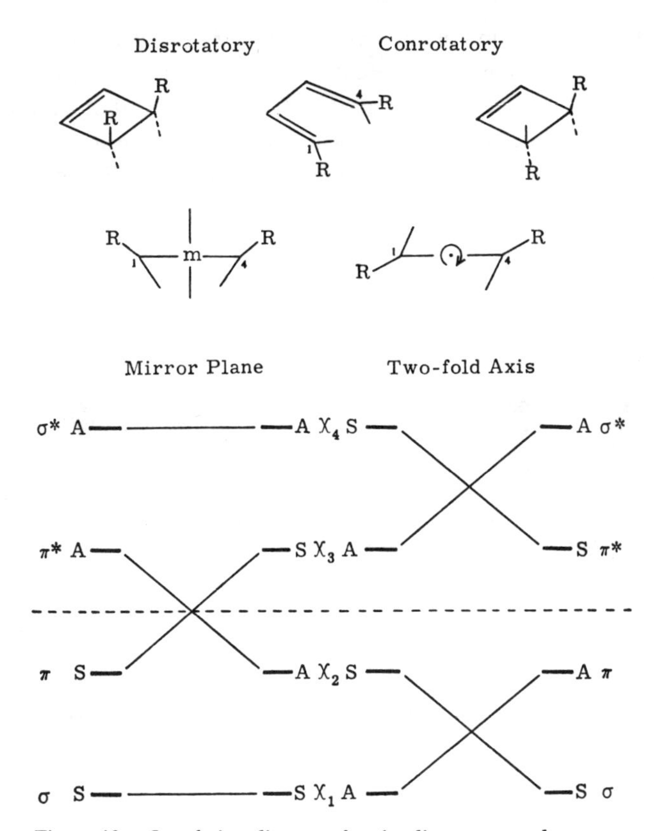

Figure 19. Correlation diagrams for the disrotatory and conrotatory conversion of cyclobutenes to butadienes.

acteristic of a symmetry-allowed reaction, while the pattern for the disrotatory process is that of a symmetry-forbidden reaction.

We should emphasize at this point that this detailed stepwise analysis has been presented primarily for pedagogic reasons. The high molecular symmetry present in these cases would have permitted very simple direct derivation of the relevant correlation diagrams. But soon we will encounter cases of such low symmetry that the stepwise analysis is the only possible one.

The highest occupied orbitals play a dominant role in these correlations. Their importance is easy to justify. First, we think of them as containing the valence electrons of the molecule, most easily perturbed during incipient reaction. In this sense their role has been stressed in the important work of *Fukui*[12]. Second, if there is little symmetry in a molecule and if there is a bonding level which is intending to cross the energy gap to correlate to an antibonding level, then that level will usually be the highest occupied level. Consequently, its motion determines the course of the correlation diagram and its initial slope is an important indication of whether the process is symmetry-allowed or forbidden. Consider the disrotatory and conrotatory motions as they affect χ_2 in butadiene *(50)*.

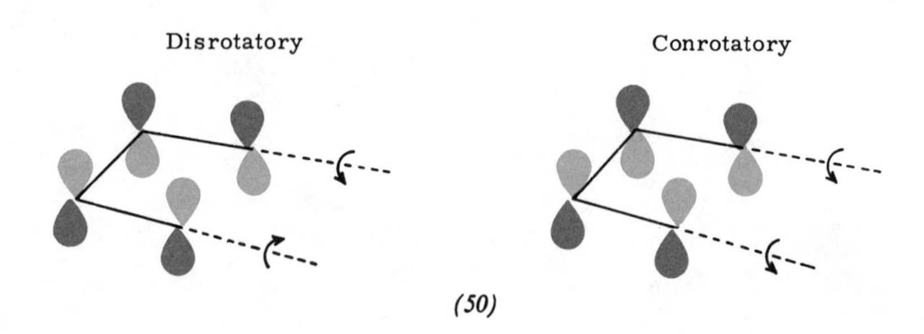

(50)

The disrotatory process pushes a plus lobe onto a minus lobe. Since one end of the molecule "feels" the phase of the wave function at the other end, this is an antibonding, destabilizing, repulsive interaction. The level moves up in energy along the reaction coordinate. Conrotatory motion brings a plus lobe onto a plus lobe (or minus on minus, which is equivalent). This is a bonding, stabilizing, attractive interaction terminating in the actual formation of the new σ bond.

A warning note should be sounded here. The sense of orbital symmetry control of any concerted reaction can always be determined through inspection of the behavior of the highest occupied molecular orbital in the reacting system, but the analysis is often less simple than it is in the case of the butadiene → cyclobutene transformation. A temptation especially to be avoided in making such analyses is that of inadvertently placing more than two electrons in a single molecular orbital.

---

[12] *K. Fukui*, *T. Yonezawa*, and *H. Shingu*, J. chem. Physics *20*, 722 (1952); *K. Fukui*, *T. Yonezawa*, *C. Nagata* and *H. Shingu*, ibid. *22*, 1433 (1954); *K. Fukui* in *O. Sinanoğlu:* Modern Quantum Chemistry. Academic Press, New York 1965, Vol. 1, p. 49, and references therein.

The general rules for electrocyclic reactions are very easily derivable from the nodal properties of polyenes and polyenyl ions: the thermal electrocyclic reactions of a k π electron system will be disrotatory for $k = 4q+2$, conrotatory for $k = 4q$ $(q = 0,1,2\ldots)$; in the first excited state these relationships are reversed. Some of the consequences of these rules are summarized in Figure 20.

Reaction	Ground State	Excited State
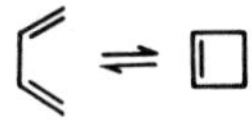	Conrotatory	Disrotatory
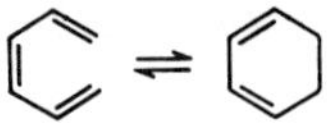	Disrotatory	Conrotatory
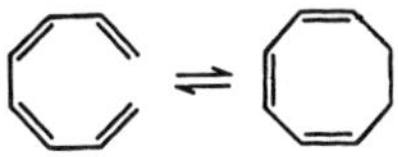	Conrotatory	Disrotatory
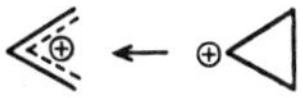	Disrotatory	Conrotatory
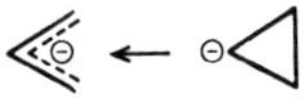	Conrotatory	Disrotatory
	Conrotatory	Disrotatory
	Disrotatory	Conrotatory

Figure 20. Conrotatory and disrotatory electrocyclic reactions.

We turn next to a consideration of secondary factors which influence the actual composition of the products of an electrocyclic reaction. The first is steric control. For every reaction there are two conrotatory and two disrotatory motions which may or may not be distinguishable. Thus, the two conrotatory modes of opening of a *cis*-dimethylcyclobutene are enantiomeric and lead to the same product, *cis,trans*-1,4-dimethylbutadiene *(51)*.

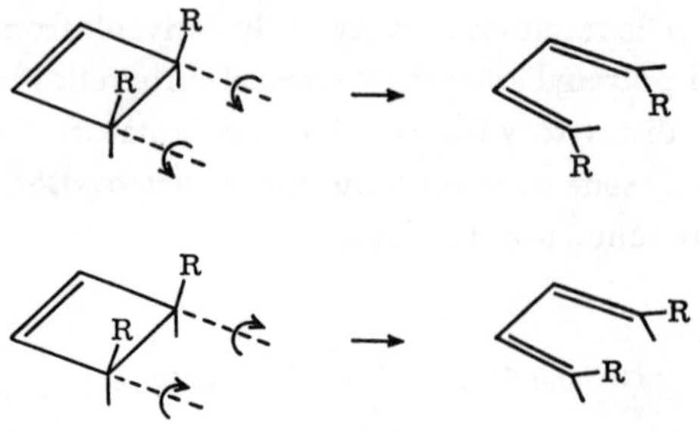

(51)

The two conrotatory motions for a *trans*-dimethylcyclobutene lead to two different isomers: *cis,cis*- and *trans,trans*-1,4-dimethylbutadiene *(52)*. In fact, the *trans*-

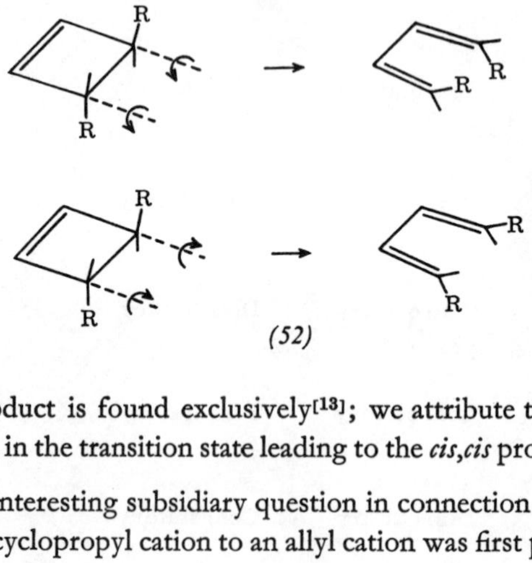

(52)

trans product is found exclusively[13]; we attribute this to the unfavorable steric situation in the transition state leading to the *cis,cis* product.

A most interesting subsidiary question in connection with the electrocyclic opening of a cyclopropyl cation to an allyl cation was first posed to us by *C.H. DePuy:* assuming that departure of a leaving group from a cyclopropane ring, and the bond-breaking electrocyclic reaction to give an allyl cation, are concerted, could there be a difference between the two *a priori* possible disrotatory modes, defined in relation to the position of the leaving group? Extended Hückel calculations provided the initial answer, which may be summarized by saying that the substituents on the same side of the three-membered ring as the leaving group rotate towards one another *(53)*, whereas those on the other side rotate apart *(54)*.

---

[13] *R. E. K. Winter*, Tetrahedron Letters *1965*, 1207.

(53)

(54)

The result may be understood in qualitative terms when it is realized that as the 2,3 bond is broken by disrotatory outward rotation *(55)*, the electron density of that bond, which originally was more or less in the plane of the cyclopropane ring, shifts above the plane. It is then available for backside displacement of the leaving group — in other words, the reaction is a normal S_N2 displacement of the group X by the electrons of the backbone σ bond of the cyclopropane ring.

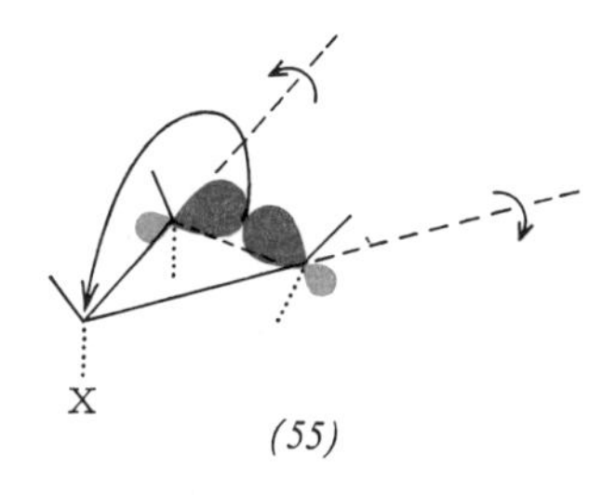

(55)

Several corollaries of these conclusions follow. If R in *(53)* or *(54)* is some bulky group then we should expect for steric reasons a faster solvolysis for compound *(54)*. On the other hand, when *cis* positions are linked by a short methylene chain, we should expect the opening of a compound such as *(56)* with leaving group *anti* to the ring to be severely disfavored, since the resulting rotation would lead to a *trans,trans*-allyl cation in a small ring. We should expect a facile opening only for a *syn* leaving group, as in *(57)*.

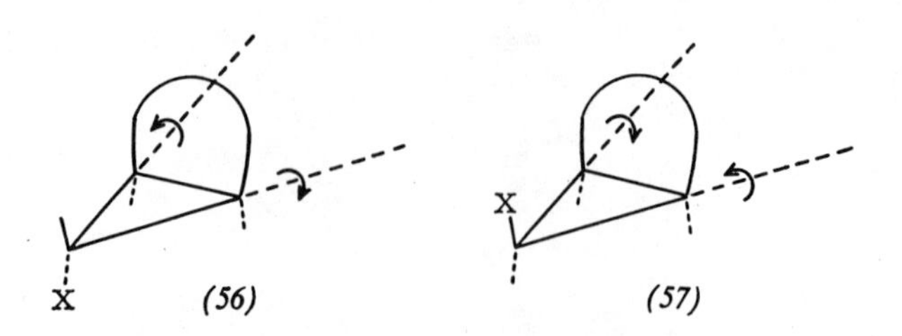

(56) (57)

5.1. Electrocyclic Reactions Exemplified

The conrotatory electrocyclic change of a cyclobutene to a butadiene is a very well-known process. Its stereochemistry was established some ten years ago[14] and its scope and energetics have been very carefully studied[15]. Satisfactory rationalization of the striking stereospecificity of the reaction was lacking until the orbital symmetry control of electrocyclic changes was discovered. Given that in the thermal reaction conrotatory displacement is symmetry-allowed, it follows that a bicyclic cyclobutene containing a *cis*-fused methylene chain *(58)* leads to a cyclic *cis-trans*-diene *(59)*.

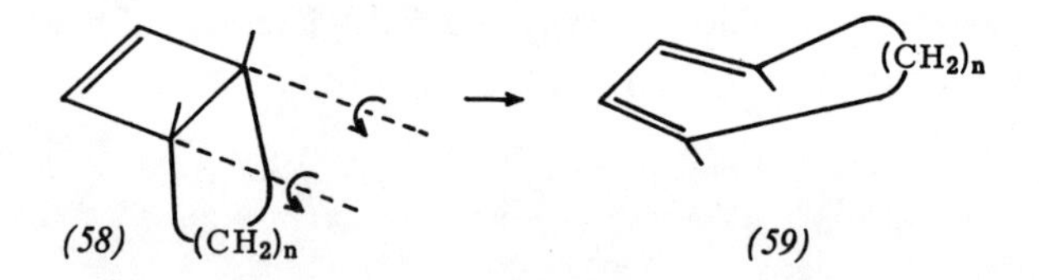

(58) (59)

Such a *trans*-olefin becomes increasingly highly strained as the ring size diminishes. This trend is beautifully reflected in the temperatures at which the half-life of compound *(58)* is about 2 hours[16] as n is varied:

n	1	2	3	4	5	6
T(°C)	<100	195	>380	350	335	180

---

[14] *E. Vogel*, Liebigs Ann. Chem. *615*, 14 (1958); *R. Criegee* and *K. Noll*, ibid. *627*, 1 (1959).
[15] *R. Criegee, D. Seebach, R. E. Winter, B. Börretzen*, and *H.-A. Brune*, Chem. Ber. *98*, 2339 (1965) and references therein; *G. R. Branton, H. M. Frey*, and *R. F. Skinner*, Trans. Faraday Soc. *62*, 1546 (1966) and references therein; *H. M. Frey* and *R. Walsh*, Chem. Rev. 69, 103 (1969).
[16] *R. Criegee* and *G. Bolz*, unpublished work; *D. Seebach*, personal communication. The value for n=1 is estimated from *J. I. Brauman, L. E. Ellis, and E. E. van Tamelen*, J. Amer. chem. Soc. *88*, 846 (1966). A more detailed study of the case *(58)*, n = 3, has been reported recently: *J. J. Bloomfield, J. S. McConaghy jr.*, and *A. G. Hortmann*, Tetrahedron Letters *1969*, 3723.

The half-life of *(58)* decreases for small *n* since, due to ring strain, the reaction itself becomes highly exothermic.

Molecules such as bicyclo[2.2.0]hexadiene and bicyclo[2.1.0]pentene clearly owe whatever stability they possess to the fact that the transformations, respectively to benzene and cyclopentadiene, are symmetry-forbidden.

Criegee and *Reinhardt*[17] have synthesized the laterally fused *cis*- and *trans*-cyclobutenes *(60)*–*(63)*. These exhibited dramatically different pyrolytic stabilities[18]. In the substances with *anti* backbones, the symmetry-imposed conrotatory motion

	E_a (kcal/mole)	$k = 10^{-4}$ at °C
(60)	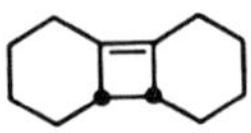 42	261
(61)	 29	87
(62)	 45	273
(63)	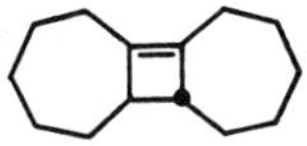 27	109

leads to *cis* double bonds in the six- or seven-membered rings and the electrocyclic cleavage takes place readily. By contrast, the symmetry-allowed processes with the *syn* isomers would give substances containing a *trans* double bond in a six- or a seven-membered ring. Consequently, the transformation occurs only at extreme temperatures; it might well proceed by a non-concerted mechanism.

The highly substituted butadienes *(64)* and *(66)* have been studied in a most ingenious and amusing experiment[19]. These substances are equilibrated through

---

[17] *R. Criegee* and *H. G. Reinhardt*, Chem. Ber. *101*, 102 (1968).

[18] Indications of a similar preference in laterally fused nine-membered rings had been found earlier: *K. G. Untch* and *D. J. Martin*, J. Amer. chem. Soc. *87*, 4501 (1965).

[19] *G. A. Doorakian* and *H. H. Freedman*, J. Amer. chem. Soc. *90*, 5310, 6896 (1968).

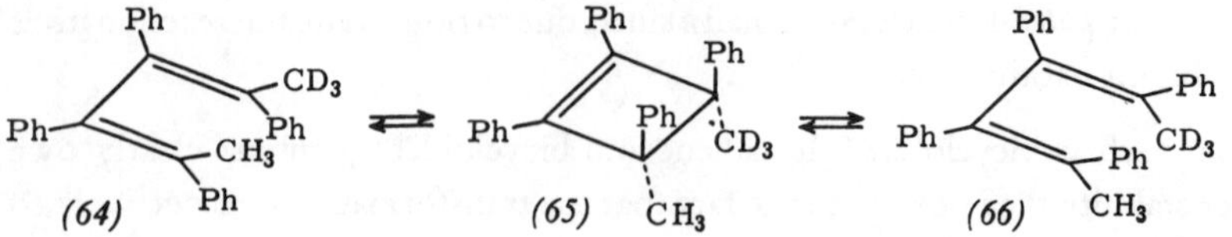

(64) (65) (66)

the intermediacy of the cyclobutene *(65)*. After fifty-one days at 124°C, each cyclobutene molecule had faultlessly undergone 2.6×10^6 conrotatory openings, and a disrotatory mistake was yet to appear.

Some fascinating transformations of the *cis*- and *trans*-benzocyclobutenes *(67)* and *(68)* are readily explicable as electrocyclic reactions followed by Diels-Alder additions[20].

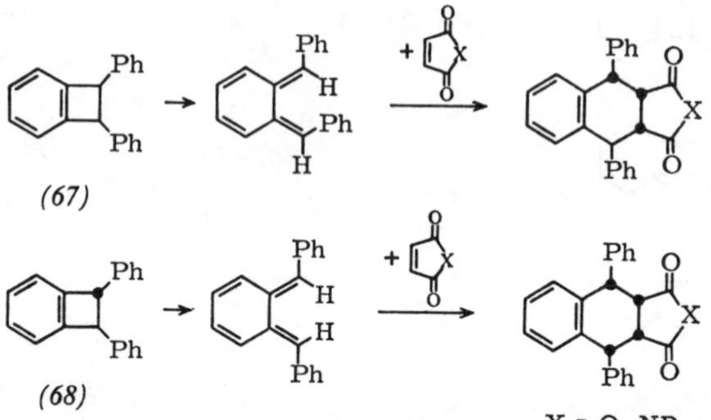

(67)

(68)

X = O, NR

There are numerous examples of photochemical cyclobutene-butadiene interconversions. In the great majority of cases the diene is part of a cyclic system and is thus constrained to undergo the symmetry-allowed disrotatory process. Some typical examples are the reactions *(69)* and *(70)*[21,22].

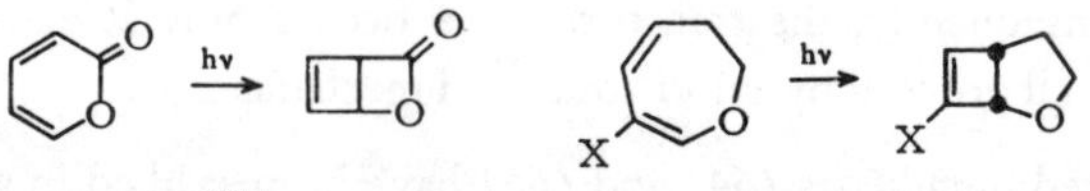

(69) X = H, Cl (70)

---

[20] *R. Huisgen* and *H. Seidel*, Tetrahedron Letters *1964*, 3381; *G. Quinkert*, *K. Opitz*, *W. W. Wiersdorff*, and *M. Finke*, ibid. *1965*, 3009.

[21] *E. J. Corey* and *J. Streith*, J. Amer. chem. Soc. *86*, 950 (1964).

[22] *L. A. Paquette*, *J. H. Barrett*, *R. P. Spitz*, and *R. Pitcher*, J. Amer. chem. Soc. *87*, 3417 (1965).

One case with greater stereochemical freedom *(71)* has been observed[23] and fits our expectations.

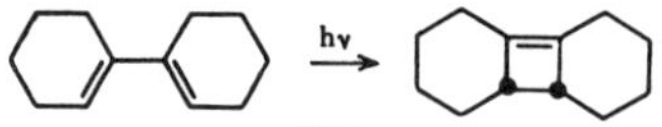

hv

(71)

Recently it has been confirmed that *trans,trans*-2,4-hexadiene undergoes photochemical cyclization to *cis*-dimethylcyclobutene[24]; the case is of special interest in that geometric constraints are absent.

The diazepinone *(72)* yields on photolysis the bicyclic isomer *(73)*, which reverts readily thermally to the starting compound[25]. The reversibility is easily understood once it is realized that direct *cis→trans* interconversion, an impossible process in fused carbocyclic systems, is simply an inversion at the nitrogen atom in *(73)*.

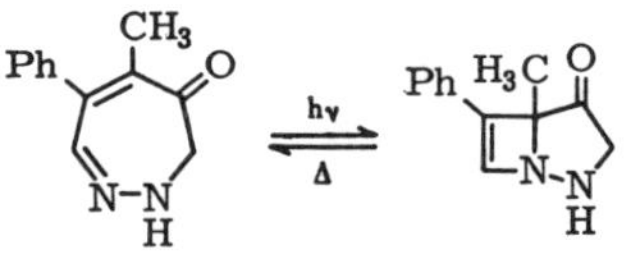

Ph CH_3 O N–N H hv Δ Ph H_3C O N–N H

(72) *(73)*

The symmetry-allowed photochemical conrotatory cyclization of *cis*-hexatrienes, and the reverse reaction, were first recognized in studies in the vitamin D field. The conversions *(74)*–*(80)* — all symmetry-allowed — were established in an extensive series of elegant investigations[26].

Of particular interest is the fact that the cyclohexadienes *(77)* and *(78)*, prevented by insurmountable geometrical restraints from undergoing symmetry-allowed

---

[23] *W. G. Dauben, R. G. Cargill, R. M. Coates,* and *J. Saltiel,* J. Amer. chem. Soc. *88,* 2742 (1966); *K. J. Crowley,* Tetrahedron *21,* 1001 (1965).
[24] *R. Srinivasan,* J. Amer. chem. Soc. *90,* 4498 (1968).
[25] *W. J. Theuer* and *J. A. Moore,* Chem. Commun. *1965,* 468.
[26] *E. Havinga, R. J. de Kock,* and *M. P. Rappold,* Tetrahedron *11,* 278 (1960); *E. Havinga* and *J. L. M. A. Schlattmann,* ibid. *16,* 146 (1961); *G. M. Sanders* and *E. Havinga,* Rec. Trav. chim. *83,* 665 (1964); *H. H. Inhoffen* and *K. Irmscher,* Fortschr. Chem. org. Naturstoffe *17,* 70 (1959); *H. H. Inhoffen,* Angew. Chem. *72,* 875 (1960); *B. Lythgoe,* Proc. chem. Soc. *1959,* 141; *W. G. Dauben* and *G. J. Fonken,* J. Amer. chem. Soc. *81,* 4060 (1959).

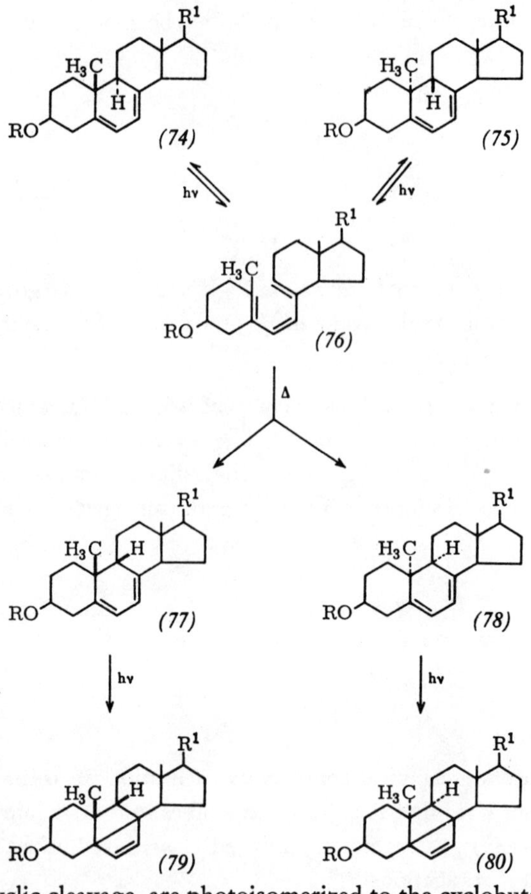

electrocyclic cleavage, are photoisomerized to the cyclobutenes *(79)* and *(80)*, in an alternative symmetry-allowed process.

A further example of the photochemical cyclohexadiene→hexatriene reaction has recently been studied[27].

A related reaction presumably also takes place in the photocyclization of *cis*-stilbenes *(81)* and similar compounds[28], though the stereochemistry of the product has not been definitely established.

---

[27] *P. Courtot* and *R. Rumin*, Tetrahedron Letters *1968*, 1091.

[28] *K. A. Muszkat* and *E. Fisher*, J. chem. Soc. *B*, *1967*, 662; *F. B. Mallory*, *C. S. Wood*, and *J. T. Gordon*, J. Amer. chem. Soc. *86*, 3094 (1964). The subject has been recently reviewed by *M. Scholz*, *F. Dietz*, and *M. Möhlstadt*, Z. Chem. *7*, 329 (1967).

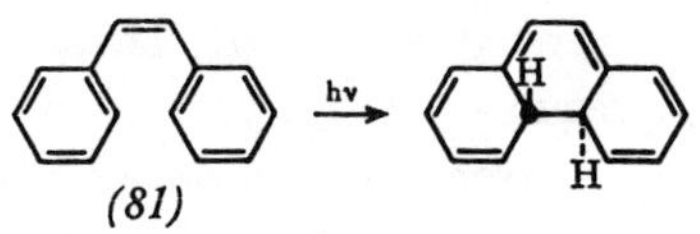

(81)

The earlier indications that the *thermal* cyclization of trienes was disrotatory[26] were confirmed by the study of simple model compounds *(82)* [29].

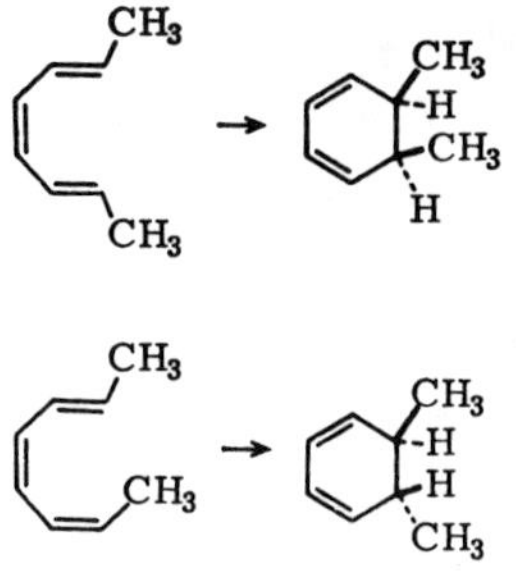

(82)

The electrocyclic closure *(83)* of a *cis,cis*-octatetraene should be thermally conrotatory, photochemically disrotatory. Though any conformation of a *cis,cis*-octatetraene is far from planarity, the same nodal patterns in the orbitals are preserved. The predicted stereochemical course of the thermal reaction has been confirmed recently[30].

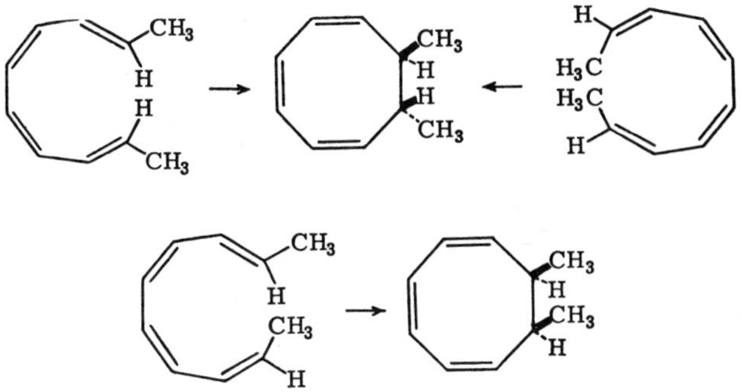

(83)

[29] *E. Vogel, W. Grimme*, and *E. Dinné*, Tetrahedron Letters *1965*, 391; *E. N. Marvell, G. Caple*, and *B. Schatz*, ibid. *1965*, 385; *D. S. Glass, J. W. H. Watthey*, and *S. Winstein*, ibid. *1965*, 377.
[30] *R. Huisgen, A. Dahmen*, and *H. Huber*, J. Amer. chem. Soc. *89*, 7130 (1967); *cf.* also *E. N. Marvell* and *J. Seubert*, ibid. *89*, 3377 (1967).

Some previous experiments can also be interpreted accordingly. Thus, in reactions *(84)* explored by *Meister*[31] the following sequence of symmetry-allowed changes can be adduced: an 8-electron conrotatory electrocyclic reaction, a 6-electron disrotatory reaction, a [4+2] cycloaddition, a reversion of the same, and a 4-electron conrotatory electrocyclic reaction.

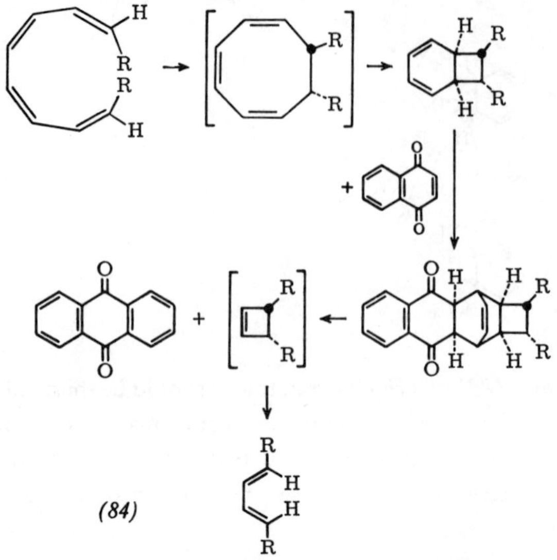

(84)

The preparation of the fantastic hydrocarbon *(85)* by *Greene*[32] presents a situation of great interest, related to the octatetraene→cyclooctatriene electrocyclic conversion. Undoubtedly, the substance would suffer instantaneous transformation to its isomer 9,9′-bianthryl *(86)* — in fact, $t_{1/2} \approx 30$ minutes at 80°C — were it not for the

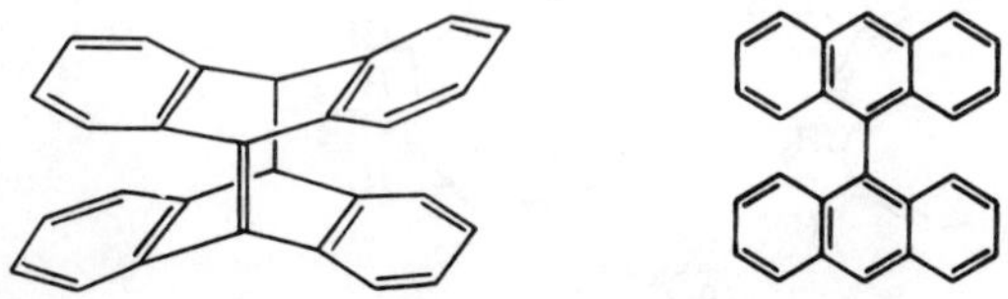

(85) *(86)*

[31] *H. Meister*, Chem. Ber. *96*, 1688 (1963).
[32] *N. M. Weinshenker* and *F. D. Greene*, J. Amer. chem. Soc. *90*, 506 (1968).

circumstances that the symmetry-allowed conrotatory transformation is opposed by a necessary concomitant, and difficult, twisting motion about the 9,9′ double bond, while the geometrically simpler disrotatory cleavage is symmetry-forbidden.

A ten-electron electrocyclic reaction may have been realized in the Ziegler-Hafner synthesis of azulene[33]. The crucial cyclization step is very probably a disrotatory electrocyclic reaction *(87)*.

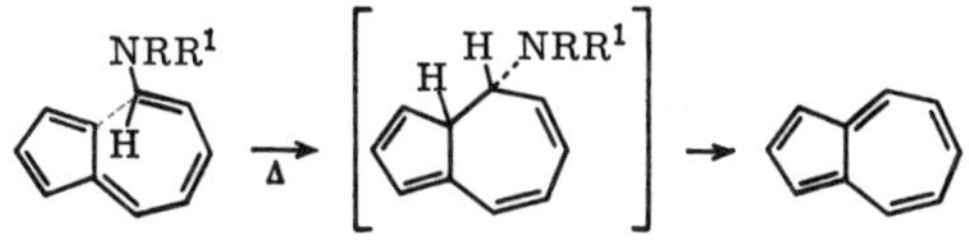

(87)

There were some previous indications in the literature that the opening of cyclopropyl cations to allyl cations is a stereospecific process[34]. For instance, *Skell* and *Sandler* observed that the two epimers *(88)* and *(89)* (structures not assigned) gave different solvolysis products. The hitherto mysterious phenomena are now

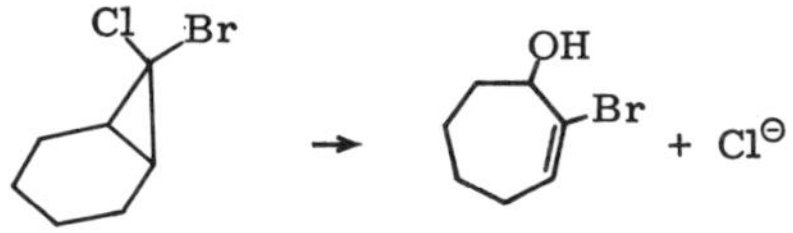

(88)

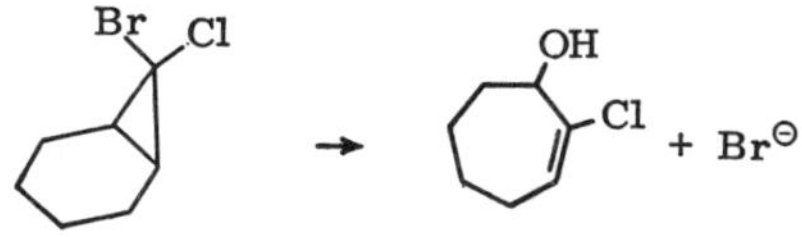

(89)

readily explicable; the structures must be assigned as shown, since only loss of the *endo* halogen atoms can be concerted with the required disrotatory opening of the cyclopropane rings.

---

[33] *K. Ziegler* and *K. Hafner*, Angew. Chem. *67*, 301 (1955); *K. Hafner*, ibid. *67*, 301 (1955).
[34] (a) *P. S. Skell* and *S. R. Sandler*, J. Amer. chem. Soc. *80*, 2024 (1958); (b) *E. E. Schweizer* and *W. E. Parham*, ibid. *82*, 4085 (1960); (c) *R. Pettit*, ibid. *82*, 1972 (1960).

The first clear confirmations of our predictions for the electrocyclic opening were obtained in 1965. Whereas *(90)* undergoes solvolysis readily at 125°C, its epimer *(91)* is recovered unchanged after prolonged treatment with acetic acid at 210°C [35].

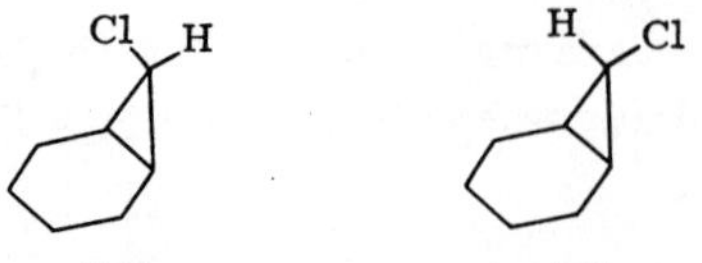

Cl H H Cl

(90) *(91)*

A detailed study of all methyl-substituted cyclopropyl tosylates clearly showed the steric effect of methyl groups forced against each other by the stereoelectronic factor[36].

Since then, numerous further confirmations have appeared. We mention here only two of these. *Ghosez* and co-workers[37] observed the reactions *(92)*, and *Whitham*

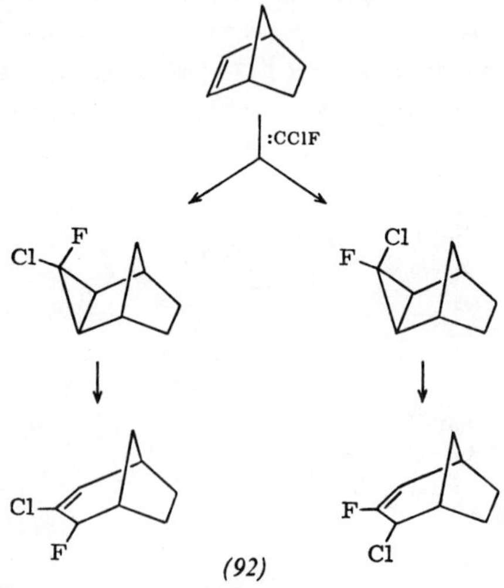

:CClF

F Cl Cl F

Cl F F Cl

(92)

---

[35] *S. F. Cristol, R. M. Segueira*, and *C. H. DePuy*, J. Amer. chem. Soc. *87*, 4007 (1965). Similar results were communicated to us by *W. Kirmse*. Cf. also the behavior on pyrolysis of the closely related chlorobicyclo[3.1.0]hexanes: *M. S. Baird* and *C. B. Reese*, Tetrahedron Letters *1967*, 1379.

[36] *P. v. R. Schleyer, G. W. Van Dine, U. Schöllkopf*, and *J. Paust*, J. Amer. chem. Soc. *88*, 2868 (1966); *U. Schöllkopf*, Angew. Chem. *80*, 603 (1968); Angew. Chem. internat. Edit. *7*, 588 (1968).

[37] *L. Ghosez, G. Slinckx, M. Glineur, P. Hoet*, and *P. Laroche*, Tetrahedron Letters *1967*, 2773. Related reactions with HCCl and HCBr are also observed: *C. W. Jefford, E. Huang Yen*, and *R. Medary*, ibid. *1966*, 6317.

has shown that solvolysis of *exo*-8-bromobicyclo[5.1.0]octane gives the expected *trans*-cyclooctenol *(93)* [38].

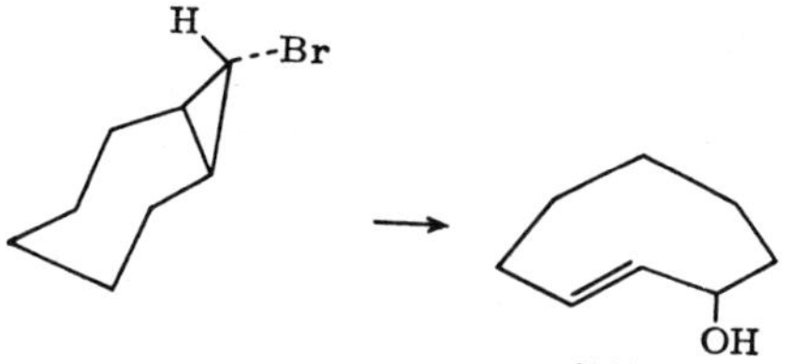
H
Br
OH

(93)

The electrocyclic cleavage of the cyclopropyl anion has not been tested directly in the parent case, but an isoelectronic example, drawn from aziridine chemistry, provides a striking confirmation of the conservation of orbital symmetry. The beautiful observations of *Huisgen*, *Scheer*, and *Huber*[39] are summarized in the diagram *(94)*.

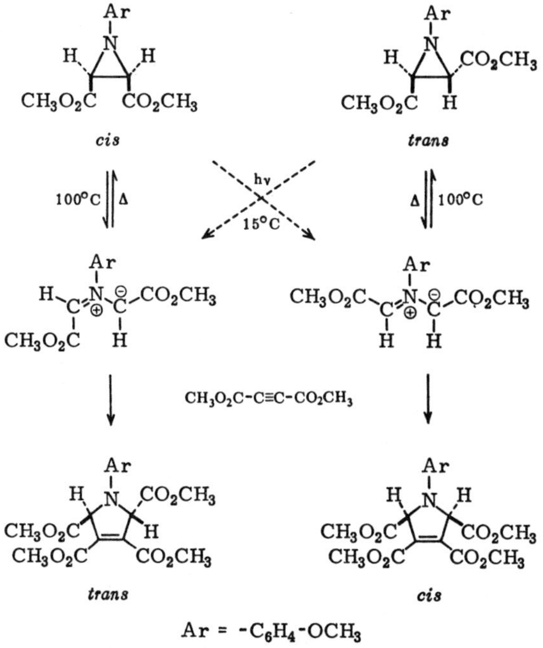
Ar
H
H
CH_3O_2C
CO_2CH_3
cis
Ar
H
CO_2CH_3
CH_3O_2C
H
trans
100°C
Δ
hv
15°C
Δ
100°C
CH_3O_2C-C≡C-CO_2CH_3
trans
cis
Ar = -C_6H_4-OCH_3

(94)

[38] *G. H. Whitham* and *M. Wright*, Chem. Commun. *1967*, 294.
[39] *R. Huisgen*, *W. Scheer*, and *H. Huber*, J. Amer. chem. Soc. *89*, 1753 (1967).

The 1,3-dipolar isomer of an aziridine is a 4 π electron molecule isoelectronic to allyl anion. The net inversion of stereochemistry observed in the thermal reaction would be extremely puzzling were it not the obvious consequence of a conrotatory opening, followed by a [4+2] cycloaddition.

The cyclization of pentadienyl cations to cyclopentenyl ions is a reaction extensively studied by *Deno* and by *Sorensen*[40]. However, the nature of the sulfuric acid medium predominantly used in these studies precludes elucidation of the stereochemistry of the reaction; hydrogen and methyl shifts intervene before product isolation. Further, it is now clear that the well-known Nazarov reaction[41] involves an electrocyclic reaction within a pentadienyl cation; again, no information about the stereochemistry of the change was forthcoming from the very extensive early work.

The stereochemistry of the reaction has now been established in recent studies at Harvard[42]. Treatment of dicyclohexenyl ketone *(95)*, R = H, with phosphoric acid affords two ketones *(96)*, R = H, and *(97)*, R = H. Similarly, the substituted ketone *(95)*,R = Me, yields *(96)*,R = Me; in this case the process is not complicated by concomitant formation of a stereochemically uninformative product of type *(97)*. Thus, the predicted conrotatory course is cleanly followed in both cases. Moreover, the irradiation of *(95)*, R = H, yields a ketone *(98)* which is the product of disrotatory cyclization.

---

[40] *N. C. Deno, C. V. Pittman, Jr.*, and *J. O. Turner*, J. Amer. chem. Soc. *87*, 2153 (1965); *T. S. Sorensen*, Canad. J. Chem. *42*, 2768 (1964); *43*, 2744 (1965); J. Amer. chem. Soc. *89*, 3782, 3794 (1967).

[41] *I. N. Nazarov* and *I. I. Zaretskaya*, Zh. Obsh. Khim. *27*, 693 (1957) and references therein.

[42] *R. Lehr, D. Kurland*, and *R. B. Woodward*, unpublished observations. *Cf. Dorothy Kurland*, Dissertation, Harvard (1967); *Roland Lehr*, Dissertation, Harvard (1968).

The opportunity for observing the excited state reactions of cations arises in mass-spectrometric studies. *Johnstone* and *Ward*[43] observed that the diphenylmethyl cation *(99)*, generated mass spectrometrically, cyclizes to a hydrofluorene species which loses *two* hydrogen atoms in a single step, and is therefore very probably the *cis* ion *(100)* — the expected product of the symmetry-allowed electrocyclization of the first excited state of the cation *(99)*. By contrast, the radical cation

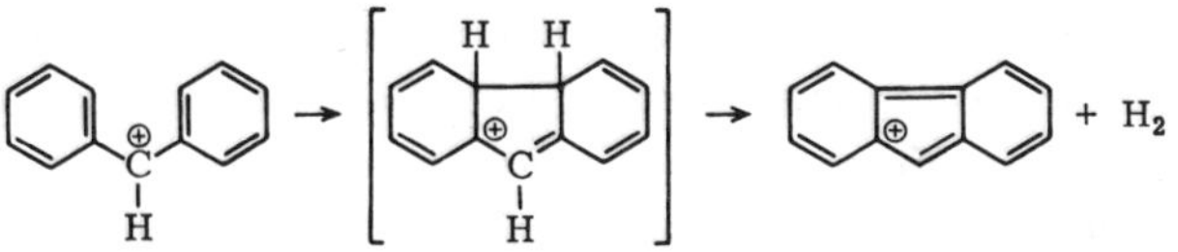

$+ H_2$

(99) *(100)*

(101), cyclizes to a hydrocarbazol species which loses hydrogen atoms one at a time, and is thus probably the *trans* ion *(102)* — again the result of the symmetry-allowed excited state process.

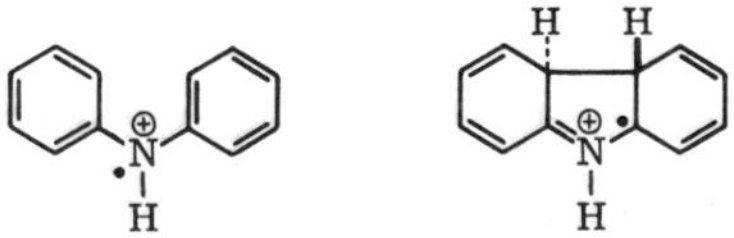

(101) *(102)*

Chapman[44] has observed photochemical cyclizations of the amines *(103)*. These are isoelectronic with pentadienyl anion, and their excited state closure should take a conrotatory course. The initially formed intermediate *(104)* apparently returns to the ground state and undergoes a hydrogen shift (a symmetry-allowed suprafacial [1,4] anionic shift — see Section 7) leading to the stable final product *(105)*.

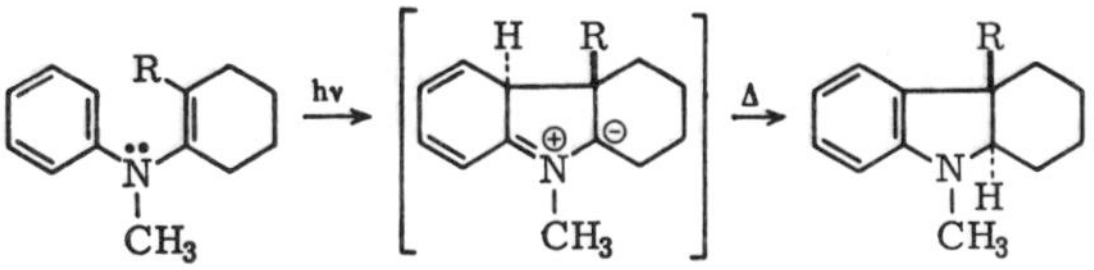

(103), R = H, CH_3 *(104)* *(105)*

---

[43] *R. A. W. Johnstone* and *S. D. Ward*, J. chem. Soc. *C, 1968*, 1805. However, *cf. M. J. Bishop* and *I. Fleming*, ibid. *1969*, 1712.
[44] *O. L. Chapman* and *G. L. Eian*, J. Amer. chem. Soc. *90*, 5329 (1968).

Very recently, the first example of the electrocyclic closure of a simple cyclopentadienyl anion has been observed [*(105a)* → *(105b)*][44a].

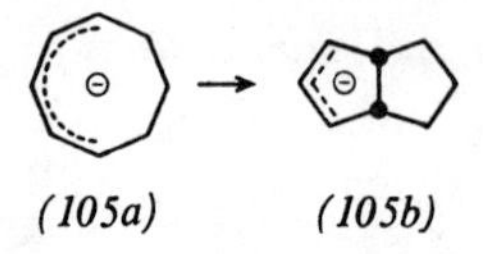

(105a) *(105b)*

Winstein has studied an exceptionally interesting system in which nine- and ten-electron electrocyclic processes are operative[45]. When one electron is added to *cis*-bicyclo[6.1.0]nona-2,4,6-triene *(106)*, an ion-radical *(107)* is produced, clearly with symmetry-allowed disrotatory geometrical displacements. Moreover, yet another electron can be added, to give a ten-electron doubly charged anion. In spec-

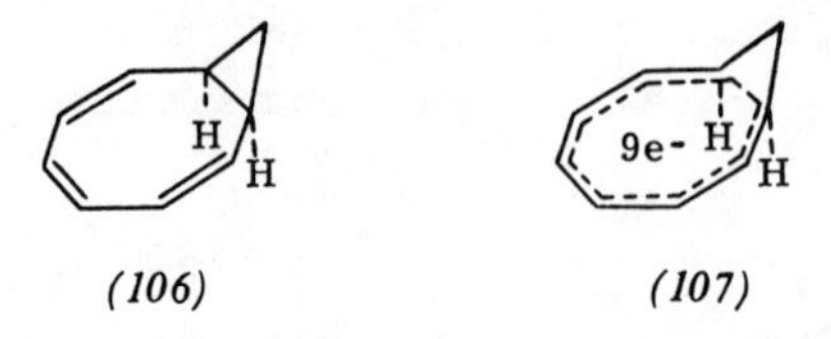

(106) *(107)*

tacular contrast[46], addition of an electron to the isomeric *trans*-triene *(108)* is not accompanied by delocalization of the cyclopropane electrons. In this case, the steric restraints present do not permit disrotatory displacements, and the seven-electron ion-radical *(109)* is produced.

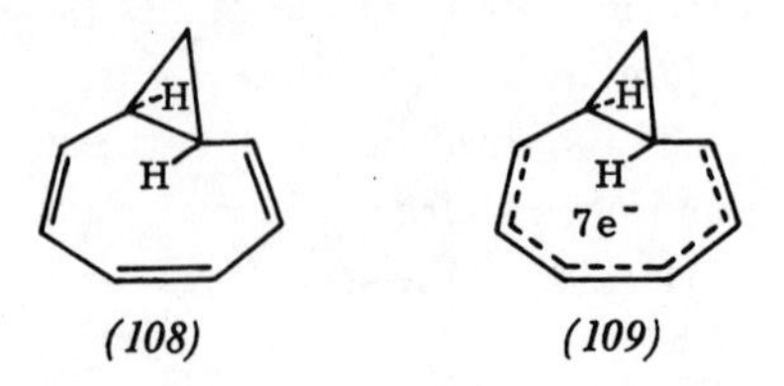

(108) *(109)*

The congeries of electrocyclic reactions *(110)*—*(116)* was studied by *Fonken*[47], who correctly assumed, but did not observe, the intermediacy of *(115)* in the

[44a] *R. B. Bates* and *D. A. McCombs*, Tetrahedron Letters *1969*, 977.
[45] *R. Rieke, M. Ogliaruso, R. McClung,* and *S. Winstein*, J. Amer. chem. Soc. *84*, 4729 (1966); *cf.* also *T. J. Katz* and *C. Talcott*, ibid. *88*, 4732 (1966).
[46] *G. Moshuk, G. Petrowski,* and *S. Winstein*, J. Amer. chem. Soc. *90*, 2179 (1968).
[47] *G. J. Fonken*, personal communication; *K. M. Schumate* and *G. J. Fonken*, J. Amer. chem. Soc. *87*, 3996 (1965); *88*, 1073 (1966).

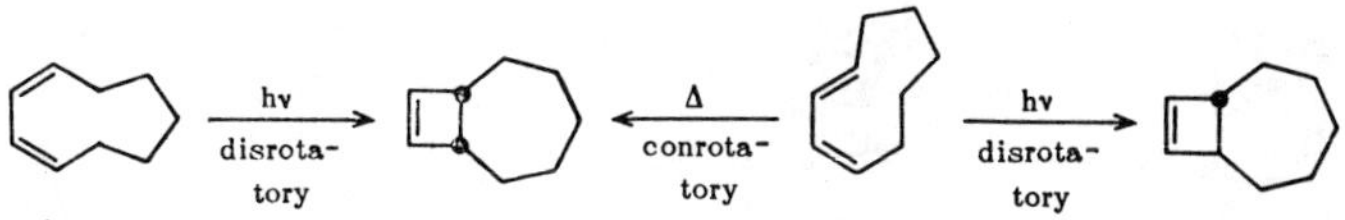

(110) —hν, disrotatory→ (111) ←Δ, conrotatory— (112) —hν, disrotatory→ (113)

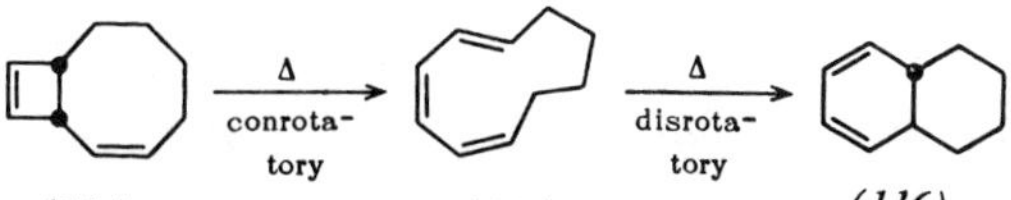

(114) —Δ, conrotatory→ (115) —Δ, disrotatory→ (116)

transformation of *(114)* into *(116)*. In an independent investigation by *Radlick*[48] this course of events has been rigorously established.

A similar sequence of electrocyclic reactions is provided by the pyrolysis of bicyclo[6.2.0]deca-2,4,6,9-tetraene to *trans*-9, 10-dihydronaphthalene *(117)* [49].

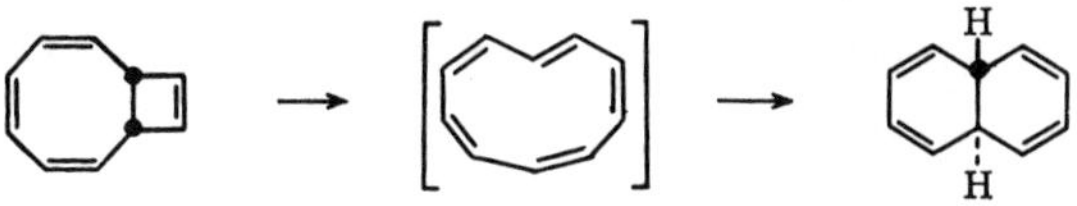

(117)

Trans-9,10-dihydronaphthalene is reported to be converted photochemically to a cyclodecapentaene (at —190°C) which reverts thermally to the *cis*-9,10-dihydronaphthalene *(118)* [50]; in more recent studies the relationships shown in the diagram have been suggested[50a].

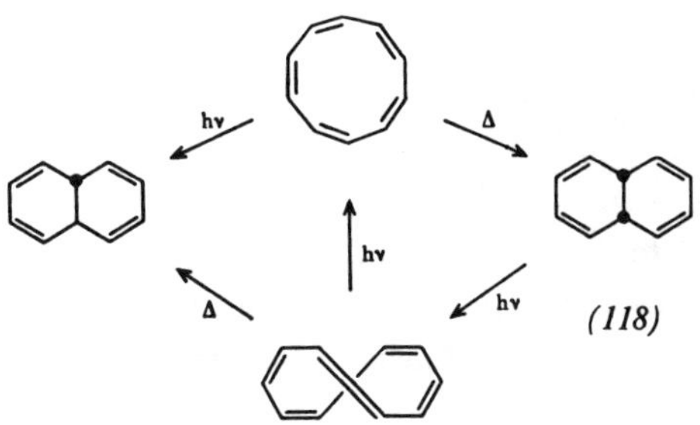

(118)

---

[48] *P. Radlick* and *W. Fenical*, Tetrahedron Letters *1967*, 4901.

[49] *S. Masamune, C. G. Chin, K. Hojo,* and *R. T. Seidner*, J. Amer. chem. Soc. *89*, 4804 (1967).

[50] *E. E. van Tamelen* and *T. L. Burkoth*, J. Amer. chem. Soc. *89*, 151 (1967). Related cases were studied by *R. C. Cookson, J. Hudec,* and *J. Marsden*, Chem. Ind. *1961*, 21, and by *E. Vogel, W. Meckel,* and *W. Grimme*, Angew. Chem. *76*, 786 (1964); Angew. Chem. internat. Edit. *3*, 643 (1964).

[50a] *S. Masamune* and *R. T. Seidner*, Chem. Commun. *1969*, 542.

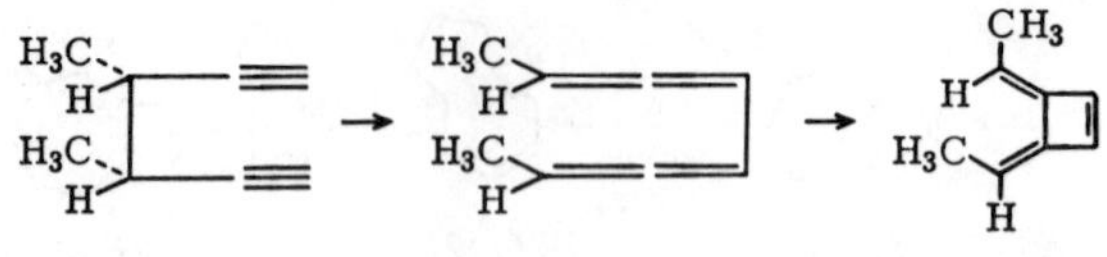

(119)

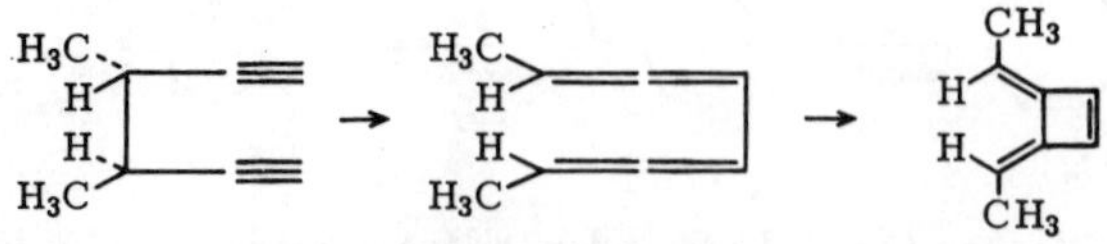

(120)

1,5-Alkadiynes undergo intramolecular rearrangement at elevated temperatures to give dimethylene cyclobutenes *(119)*, *(120)*. The observed stereospecificity[51] is readily accounted for by the sequence of a symmetry-allowed [3,3] sigmatropic shift (see below, Section 7) and a four-electron conrotatory electrocyclic reaction.

The recently synthesized [16]annulene isomerizes thermally and photochemically to two different isomers[52], the products of double disrotatory and double conrotatory closures, respectively.

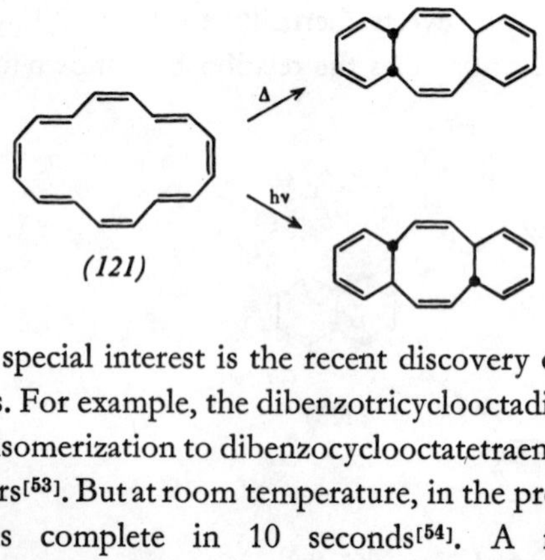

(121)

Of very special interest is the recent discovery of metal catalysis of electrocyclic reactions. For example, the dibenzotricyclooctadiene *(122)* is reported to undergo thermal isomerization to dibenzocyclooctatetraene *(124)* when heated at 180°C for 4—5 hours[53]. But at room temperature, in the presence of silver ions, the isomerization is complete in 10 seconds[54]. A forbidden disrotatory opening

[51] *W. D. Huntsman* and *H. J. Wristers*, J. Amer. chem. Soc. *89*, 342 (1967).
[52] *G. Schröder*, *W. Martin*, and *J. F. M. Oth*, Angew. Chem. *79*, 861 (1967); Angew. Chem. internat. Edit. *6*, 870 (1967).
[53] *M. Avram*, *D. Dinu*, *G. Mateescu*, and *C. D. Nenitzescu*, Chem. Ber. *93*, 1789 (1960).
[54] *W. Merk* and *R. Pettit*, J. Amer. chem. Soc. *89*, 4788 (1967).

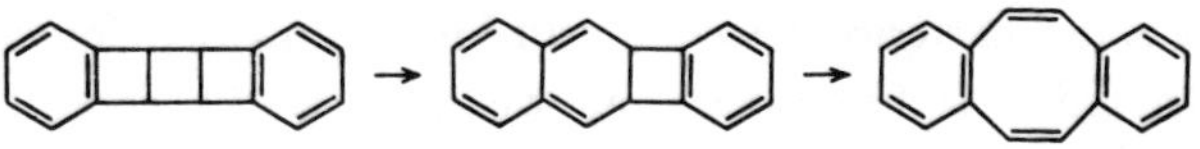

(122) (123) (124)

[*(122)*→*(123)*] is here obviously made allowed by the extra orbitals and electrons available from the complexing metal ion. A similar dramatic reversal of the rules for cycloaddition and cycloreversion reactions has been observed, and analyzed in terms of orbital symmetry conservation[55].

We now conclude our exemplification of electrocyclic reactions by alluding to two interesting, but as yet unrealized, possibilities. The valence tautomerism of cyclooctatetraene and bicyclo[4.2.0]octatriene is familiar[56]. The requisite disrotatory process may be achieved thermally by an opening of the six-membered ring of *(126)* or photochemically by cleavage of the cyclobutene ring. Since these electro-

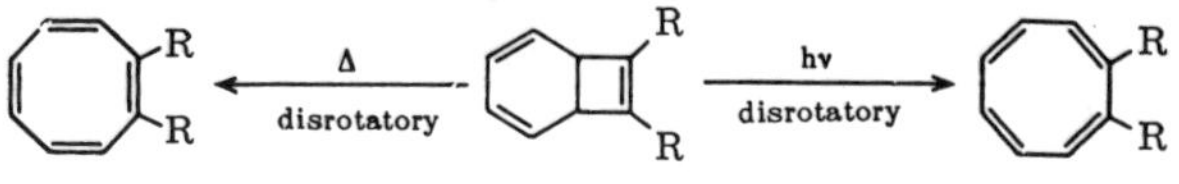

Δ disrotatory hν disrotatory

(125) (126) (127)

cyclic changes might be faster than bond switching in cyclooctatetraenes[57], it is possible that the alternatives might be experimentally observable.

(128) (129)

The as yet unsynthesized cyclodeca-1,2,4,6,7,9-hexaene *(128)* can exist in *meso* and *d,l* modifications. Both stereoisomers should be nearly strainless. The molecules are of course, valence tautomers of naphthalene *(129)*, and closure to naphthalene is formally an electrocyclic conversion of a hexatriene to a cyclohexadiene. The

[55] *F. D. Mango* and *J. H. Schachtschneider*, J. Amer. chem. Soc. *89*, 2484 (1967); *H. Hogeveen* and *H. C. Volger*, ibid. *89*, 2486 (1967).
[56] *E. Vogel, H. Kiefer*, and *W. R. Roth*, Angew. Chem. *76*, 432 (1964); Angew. Chem. internat. Edit. *3*, 442 (1964); *R. Huisgen* and *F. Mietzsch*, Angew. Chem. *76*, 36 (1964); Angew. Chem. internat. Edit. *3*, 83 (1964).
[57] *F. A. L. Anet, A. J. R. Bourn*, and *Y. S. Lin*, J. Amer. chem. Soc. *86*, 3576 (1964).

meso form *(130)* is in fact sterically well-disposed for disrotatory closure to naphthalene; since that indeed is the symmetry-allowed electrocyclic mode, it would be expected that *(130)* would have a very fragile grasp on existence. By contrast, the molecules of the *d,l* form [*cf.* *(131)*] have two interesting features. First, a model

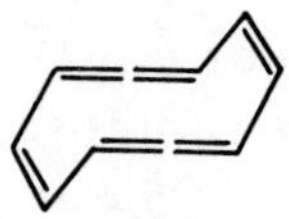

(130)  *(131)*

shows that an interesting conformational flipping can take place, without change of chirality. Second, the geometric circumstances are such that isomerization to naphthalene can be completed only by a symmetry-forbidden conrotatory closure. Consequently, the racemic compound might have reasonable thermal (but not photochemical) stability with respect to transformation into its aromatic isomer. By the same token, *(131)* is the product of the conrotatory cleavage of the 9,10 bond of the naphthalene molecule, and might well be among the products of photochemical transformation of the latter.

6. Theory of Cycloadditions and Cycloreversions

In our discussion of correlation diagrams we have already derived the selection rules for one type of simple two-component cycloaddition, namely, that which is *suprafacial* on each component. A suprafacial process is one in which bonds made or broken lie on the same face of the system undergoing reaction. For example, in an ethylenic *(132)* or cisoid diene system *(133)*, formation of bonds in the senses indicated by the arrows takes place in a suprafacial manner.

(132) *(133)*

A priori however, there are alternative, *antarafacial* processes, in which the newly formed or broken bonds lie on opposite faces of the reacting systems[58] [*cf. (134)* and *(135)*].

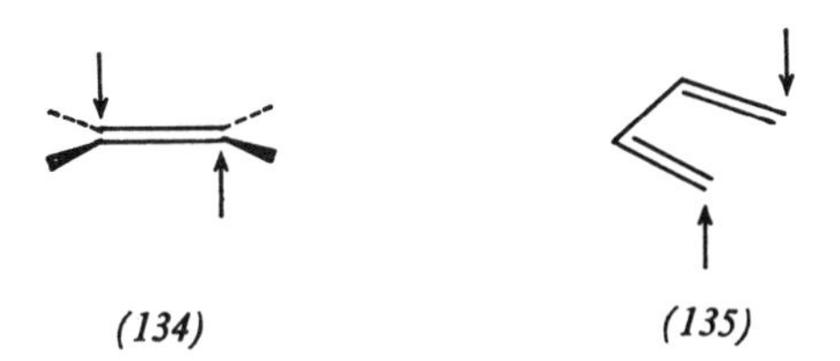

(134) *(135)*

We shall use the terms *supra* and *antara* to designate these geometrical alternatives in making general allusions to reaction types, and the letters s and a as subscripts in defining particular reactions. Thus, the cycloadditions mentioned above are *supra,supra* reactions, and the Diels-Alder reaction is a $[4_s+2_s]$ process[59].

---

[58] The terms suprafacial and antarafacial were first used in our discussion of sigmatropic reactions; *cf.* ref. [3].

[59] We [Accounts of Chem. Res. *1*, 17 (1968)] and others have earlier used *cis* and *trans* to designate the geometrical relationships here denominated *supra* and *antara*. However, the use of *cis* and *trans* as nomenclatural qualifiers is firmly established, and their employment in different senses — often of necessity, simultaneously — can be cumbersome and confusing, especially in discussing multicomponent combinations.

A priori, there are in fact four possible modes of combination of the termini of two unlike components in a cycloaddition reaction. Each mode has a characteristic stereochemical consequence, displayed in Figure 21.

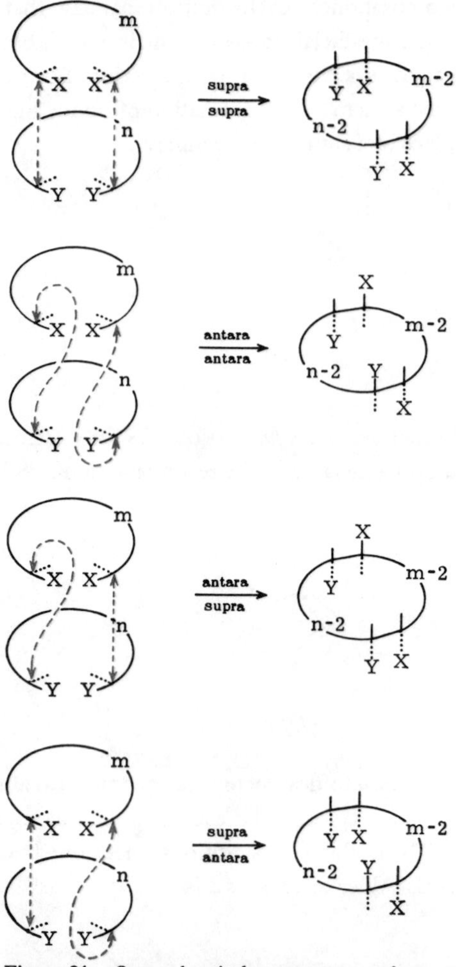

Figure 21. Stereochemical consequences of two-component cycloadditions. The diagrams are purely schematic, and do not reflect the actual geometry of the transition states, which of course differs markedly from case to case. Note that there are two possible *supra,supra* cases when m or $n > 2$, differing in the mode of approach of the reactants [*exo* and *endo* additions]. Similar circumstances obtain in the other cases. When the two reactants are identical the *supra,antara* and the *antara,supra* processes are indistinguishable.

In using the principle of orbital symmetry conservation to determine whether any cycloaddition is symmetry-allowed or forbidden, a complete analysis requires that all orbitals — bonding and antibonding — be considered. But a simplified procedure is often useful. First, all relevant reactant electrons are placed in fully delocalized bonding σ molecular orbitals of the *product* of the cycloaddition under examination. Then, if the electrons occupying the σ orbitals can be moved into bonding orbitals of the products of the cycloreversion when the σ bonds are broken with symmetry conservation, the reaction is symmetry-allowed. When the system is one lacking any molecular symmetry, the signs of the orbital lobes are determined by point-to-point transference from an analogous symmetrical system involving the same numbers of orbitals of the same types.

We shall illustrate the procedure for all possible [2+2] reactions. In each case the four relevant electrons are first placed in pairs in the two generalized cyclobutane orbitals *(136)* and *(137)*.

(136) *(137)*

a) The $[2_s+2_s]$ process. When the σ bonds are cleaved, the two electrons in the orbital *(136)* can pass with symmetry conservation into a bonding orbital *(138)* of

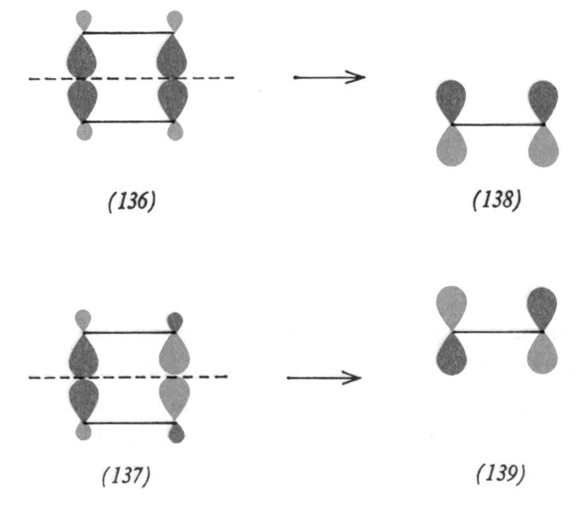

(136) *(138)*

(137) *(139)*

one ethylene molecule, but those in *(137)* can only enter an antibonding orbital *(139)* of the other ethylene.

The reaction is symmetry-forbidden.

b) The $[2_s+2_a]$ process. When the σ bonds are cleaved, the two electrons in the orbital *(136)*, now shown distorted in a purely formal manner, can pass with symmetry conservation into a bonding orbital *(140)* of one ethylene molecule, while those in *(137)* likewise pass into a bonding orbital *(141)* of the other ethylene.

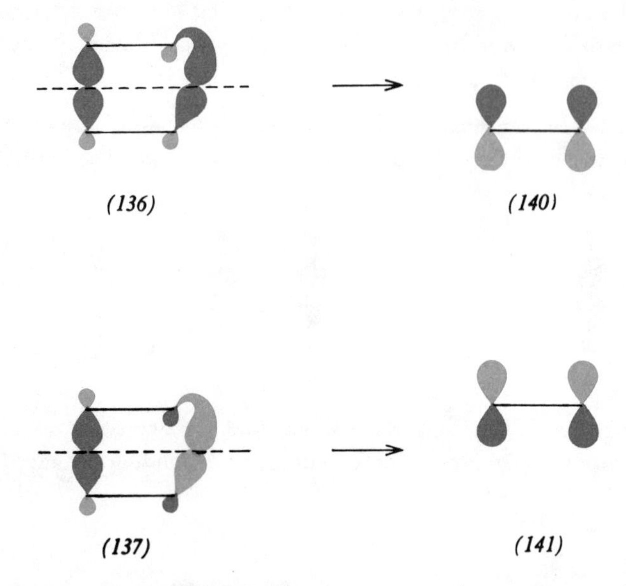

(136) (140)

(137) (141)

The reaction is symmetry-allowed.

c) The $[2_a+2_a]$ process. When the σ bonds are cleaved, two electrons in the orbital *(136)*, again, but differently, distorted in a purely formal manner, can only enter an antibonding orbital *(142)* of one ethylene molecule, while those of *(137)* pass into a bonding orbital *(143)* of the other.

The reaction is symmetry-forbidden.

It will be noted that the analyses just completed reveal the existence of a concerted symmetry-allowed path for the combination of two ethylenic compounds to give a cyclobutane. What the simplified formal analysis does not render obvious is the actual geometry of approach for the allowed $[2_s+2_a]$ process, which clearly must differ from that of the symmetry-forbidden combinations. When one considers

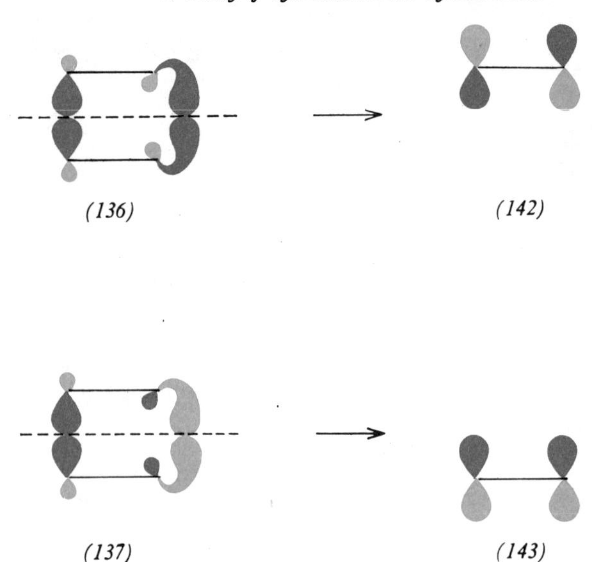
(136) (142)

(137) (143)

in detail the manner in which maximum overlap of the relevant orbital lobes may be achieved in the $[2_s+2_a]$ reaction, it is clear that the ethylenic components must approach one another orthogonally, as in *(144)*.

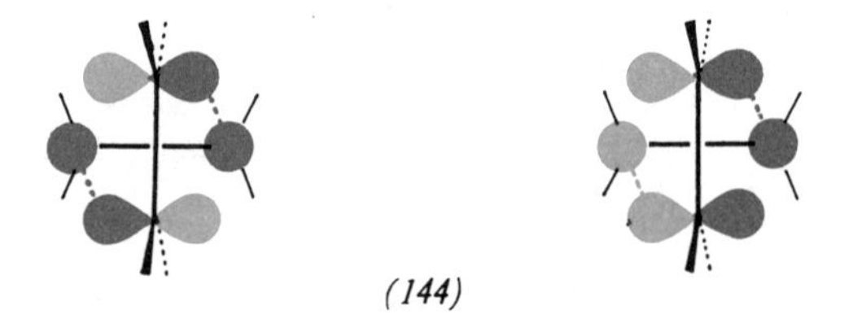
(144)

It is further worthy of comment that a complete correlation diagram may be constructed for the allowed $[2_s+2_a]$ reaction, using a two-fold axis of symmetry which passes through the midpoints of both of the ethylenic bonds. Similar circumstances obtain in the case of the $[2_a+2_a]$ reaction. In this case the geometry of approach is that shown in *(145)*, and the reaction is forbidden in the ground state, as deduced above, but allowed for the excited state (*cf.* Figure 22). A final point of much importance is that the principle of conservation of orbital symmetry is valid, *whether or not a formal correlation diagram can be constructed.* For example, no pertinent correlation diagrams can be drawn for the concerted $[4_s+2_a]$ and $[6_s+2_a]$ combi-

(145)

nations, but orbital symmetry analysis readily reveals that the former is forbidden and the latter allowed.

The generalized cycloaddition rules are now easily derived. In the addition of an m to an n-electron system the rules shown in Figure 22 must be observed for concerted processes (q is an integer $= 1, 2, 3 \ldots$).

$m+n$	Allowed in Ground State Forbidden in Excited State	Allowed in Excited State Forbidden in Ground State
$4q$	$m_s + n_a$	$m_s + n_s$
	$m_a + n_s$	$m_a + n_a$
$4q + 2$	$m_s + n_s$	$m_s + n_a$
	$m_a + n_a$	$m_a + n_s$

Figure 22. Selection rules for $[m+n]$ cycloadditions.

An important generalization is that these rules depend not on the total number of orbitals but on the number of electrons. Thus, the [4+2] case may be achieved in each of the ways shown by formulae *(146)*—*(149)*.

It should be noted also that *supra,antara* and *antara,antara* cycloadditions are not as unlikely as they might appear to be on simple steric grounds.

It remains to point out that *σ bonds can act as components in cycloaddition reactions.* In such cases, the following definitions are important:

a) A σ bond is considered to be involved in a cycloaddition reaction in a *supra* sense if configuration is retained, or inverted, at *both* of its termini in the course of reaction.

b) A σ bond is considered to be involved in a cycloaddition reaction in an *antara* sense if configuration is retained at one, and inverted at the other, terminus in the course of reaction.

(146)

(147)

(148)

(149)

The rationale underlying these important conventions will be apparent immediately upon consideration of additions to an ethylene molecule, regarded as containing two σ bonds *(150)*.

c) We adopt the simple and useful device of denominating the type of orbital involved in a concerted reaction by a subscript preceding the number of electrons in the relevant orbital. Thus

$_{\sigma}2_{a}$

$_{\pi}4_{s}$ *etc.*

In order to clarify and exemplify these matters, it may be pointed out that the electrocyclic transformation of cyclobutenes to butadienes may be regarded formally as a cycloaddition of a σ bond to a π bond, in either of two equivalent ways.

a) As shown in Figure 23: *suprafacial* on the σ bond, *antarafacial* on the π bond, *i. e.* $[_{\sigma}2_{s}+_{\pi}2_{a}]$.

b) As shown in Figure 24: *antarafacial* on the σ bond, *suprafacial* on the π bond, *i. e.* $[_{\pi}2_{s}+_{\sigma}2_{a}]$.

It will be observed that whichever formal view is taken, the conrotatory processes represent symmetry-allowed cycloaddition reactions; by contrast, the disrotatory analogues must be regarded as $[_{\pi}2_{s}+_{\sigma}2_{s}]$ or $[_{\pi}2_{a}+_{\sigma}2_{a}]$ reactions, which are, of course, symmetry-forbidden in ground states.

Suprafacial Additions

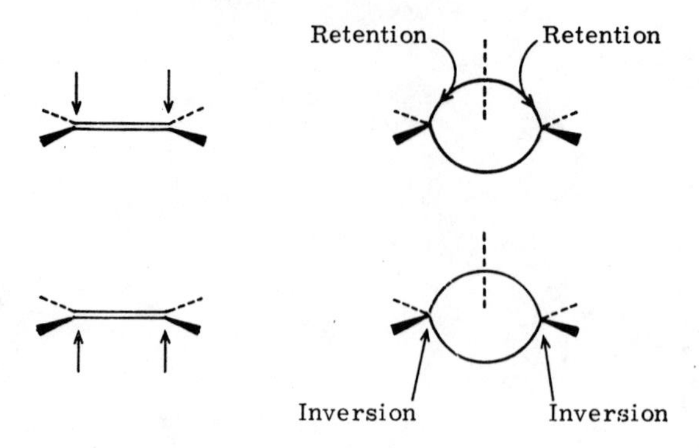

Antarafacial Additions

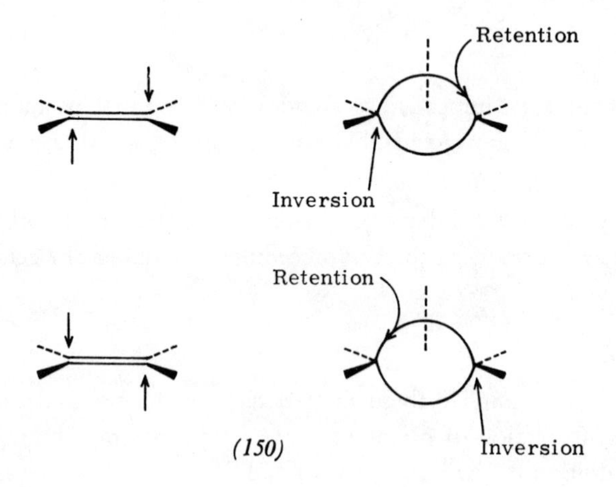

(150)

As a final illustration we may now characterize the concerted scission of cyclobutane as a $[_{\sigma}2_s + _{\sigma}2_a]$ process *(151)*.

(151)

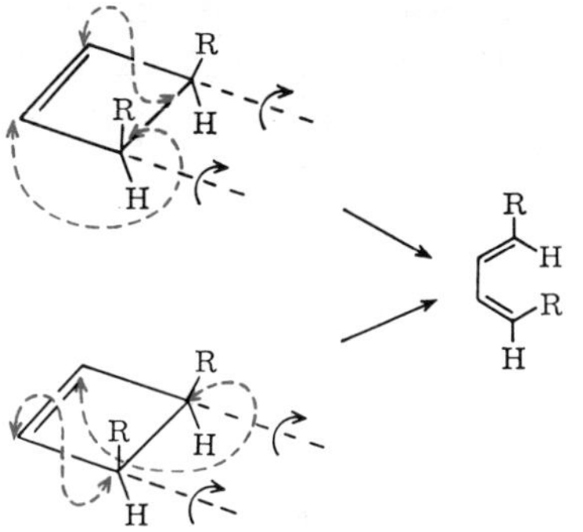

Figure 23. Electrocyclic conversion of cyclobutenes to butadienes as [${}_{\sigma}2_s + {}_{\pi}2_a$] cycloadditions.

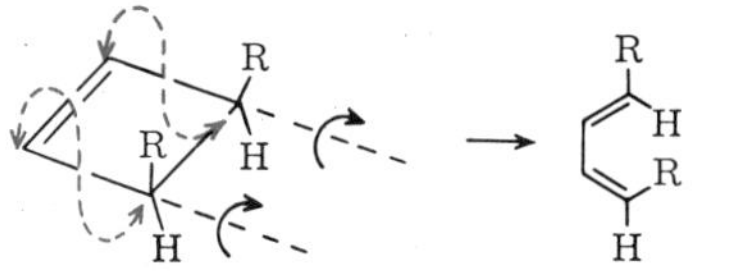

Figure 24. Electrocyclic conversion of cyclobutenes to butadienes as [${}_{\pi}2_s + {}_{\pi}2_a$] cycloadditions.

6.1. Cycloadditions and Cycloreversions Exemplified

The [2+2] cycloaddition is one of the most widely observed photochemical reactions, with more than one hundred examples quoted in a recent review[60]. It is likely that many of these reactions are not concerted, probably as a consequence of the fact that competitive relaxation of a participant excited state occurs, with transformation to an equilibrium geometry in which the ethylene moieties are 90° out of coplanarity. Thus, *trans* fusions are common in many of the cycloadducts[61]; of course, it should be noted that *trans* fused products would result from concerted symmetry-allowed [${}_{\pi}2_s + {}_{\pi}2_a$] combinations of vibrationally excited ground-state

[60] *R. N. Warrener* and *J. B. Bremner*, Rev. pure appl. Chem. *16*, 117 (1966). See also *W. L. Dilling*, Chem. Rev. *66*, 373 (1966).
[61] *E. J. Corey*, *J. D. Bass*, *R. LeMahieu*, and *R. B. Mitra*, J. Amer. chem. Soc. *86*, 5570 (1964); *P. E. Eaton* and *K. Lin*, ibid. *86*, 2087 (1964); *A. Cox*, *P. de Mayo*, and *R. W. Yip*, ibid. *88*, 1043 (1966); *R. Robson*, *P. W. Grubb*, and *J. A. Barltrop*, J. chem. Soc. *1964*, 2153.

molecules. Nevertheless, two striking examples of concerted symmetry-allowed photo-induced $[2_s+2_s]$ processes have recently been observed. Thus, irradiation of the neat *cis*- and *trans*-2-butenes, separately and in admixture, gives the results shown in *(151a)*[61a].

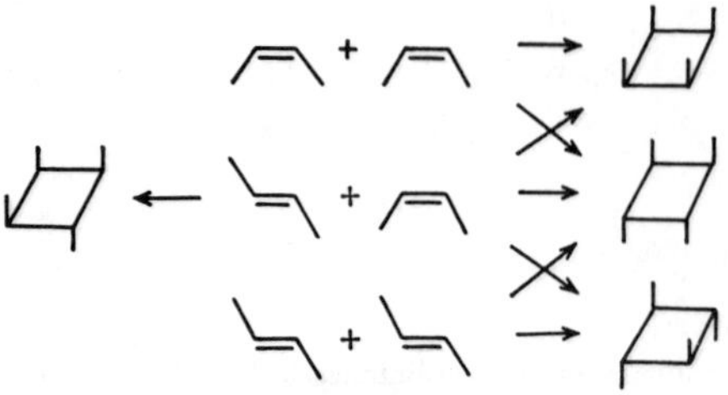

(151a)

Further, a pretty example *(151b)* of the photochemical $[_\sigma 2_s+_\sigma 2_s]$ reaction has been discovered[61b].

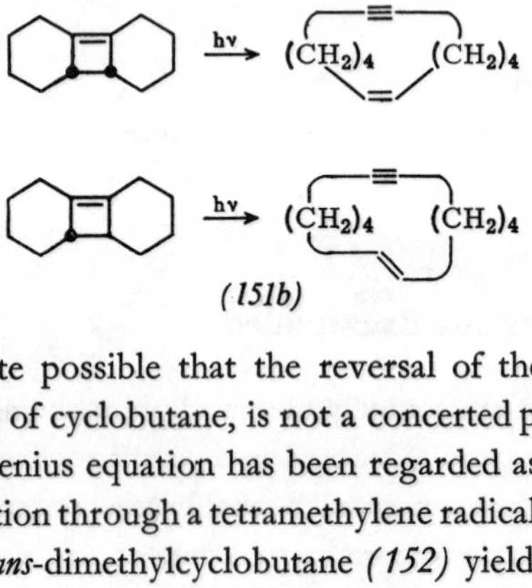

(151b)

It is quite possible that the reversal of the simplest [2+2] cycloaddition, the pyrolysis of cyclobutane, is not a concerted process. The pre-exponential factor of the Arrhenius equation has been regarded as consistent only with a stepwise decomposition through a tetramethylene radical[62]. Further, the fact that pyrolysis of *cis*- or *trans*-dimethylcyclobutane *(152)* yields a mixture of *cis*- and *trans*-butenes, among other products[63], has been adduced in support of the diradical mechanism. Nevertheless, if the cycloreversion is concerted, it must be of the type $[_\sigma 2_s+_\sigma 2_a]$, and the same stereochemical result would be observed.

[61a] *H. Yamazaki* and *R. J. Cvetanović*, J. Amer. chem. Soc. *91*, 520 (1969).
[61b] *J. Saltiel* and *L.-S. Ng Lim*, J. Amer. chem. Soc. *91*, 5404 (1969).
[62] *S. W. Benson* and *P. S. Nangia*, J. chem. Physics *38*, 18 (1963); see also the review by *H. M. Frey* in *V. Gold:* Advances in Physical Organic Chemistry. Academic Press, New York 1966, Vol. 4, p. 170.
[63] *H. R. Gerberich* and *W. D. Walters*, J. Amer. chem. Soc. *83*, 3935, 4884 (1961).

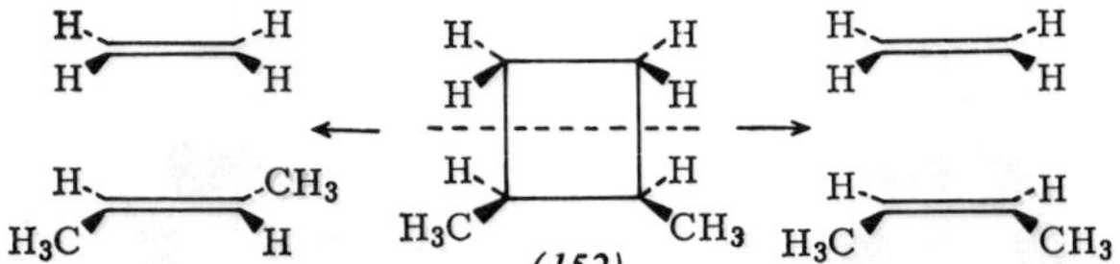

(152)

A beautiful example of the symmetry-allowed $[\pi 2_s + \pi 2_a]$ combination has been discovered by *Kraft* and *Koltzenburg*[64]. The olefin *(153)* dimerizes spontaneously to *(154)*.

(153) 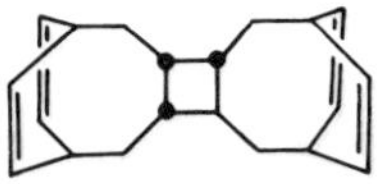 (154)

Earlier observations[65] that *cis,trans*-cycloocta-1,5-diene *(155)* dimerizes spontaneously at room temperature to a cyclobutane are less illuminating, since the stereochemistry of the product — now predictable as *(156)* — has not been established.

(155) 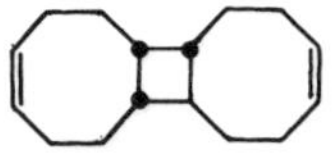 (156)

It is worthy of note that when a double bond is twisted about its axis — as it must be in *(153)* and *(155)* — the concomitant orbital twisting is such as distinctly to favor the $[\pi 2_s + \pi 2_a]$ process.

A further very interesting point which emerges from a detailed consideration of the general case (Figure 25) is that the reaction-facilitating twisting motions generate systems of opposite chirality in the two ethylene components. Consequently, *we may predict that the optically active forms of the olefins (153) and (155) will dimerize less readily than the racemic substances.*

Bicyclobutanes are produced on irradiation of some substituted butadienes[66] as well as butadiene itself[67]. The bicyclobutanes are remarkably stable, considering

[64] *K. Kraft* and *G. Koltzenburg*, Tetrahedron Letters *1967*, 4357, 4723.
[65] *K. Ziegler*, *H. Sauer*, *L. Bruns*, *H. Froitzheim-Kühlhorn*, and *J. Schneider*, Liebigs Ann. Chem. *589*, 122 (1954); *K. Ziegler* and *H. Wilms*, ibid. *567*, 1 (1950); *A. C. Cope*, *C. F. Howell*, and *A. Knowles*, J. Amer. chem. Soc. *84*, 3190 (1962).
[66] *W. G. Dauben* and *W. T. Wipke*, Pure appl. Chem. *9*, 539 (1964), and references therein.
[67] *R. Srinivasan*, *A. A. Levi*, and *I. Haller*, J. phys. Chem. *69*, 1775 (1965).

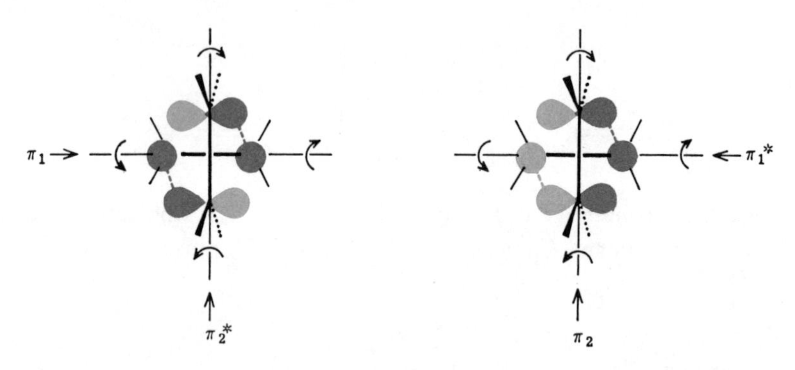

Figure 25. Combination of systems of opposite chirality in the $[_{\pi}2_s + _{\pi}2_a]$ cycloaddition of two ethylene molecules.

their strain energy (*ca.* 69 kcal/mole). The activation energy for isomerization to butadiene is 41 kcal/mole[68, 69]. The temptation arises then to consider the rearrangement of bicyclobutane to butadiene as a non-concerted process proceeding through a diradical intermediate *(157)*.

(157)

If, however, the reaction is a concerted one, it is clear that it must be a $[_{\sigma}2_s + _{\sigma}2_a]$ process, with the stereochemical consequences shown in *(158)* [70].

(158)

The evidence is now compelling that the reaction is in fact a concerted $[_{\sigma}2_s + _{\sigma}2_a]$ process. A first, indirect indication that such is the case, was provided by the observation that *(159)* is converted to *(161)* on pyrolysis[71]. The result is best accommo-

[68] *H. M. Frey* and *I. D. R. Stevens*, Trans. Faraday Soc. *61*, 90 (1965).
[69] *R. Srinivasan*, J. Amer. chem. Soc. *85*, 4045 (1963).
[70] Similar conclusions have been reached by *Wiberg* on the basis of semi-empirical MO calculations [*K. B. Wiberg*, Tetrahedron *24*, 1083 (1968)].
[71] *K. B. Wiberg* and *G. Szeimies*, Tetrahedron Letter, *1968*, 1235.

(159) (160) (161)

dated by assuming the intermediacy of *(160)*, which closes in a conrotatory fashion to the observed product. The decisive experiment was recently reported by *Closs* and *Pfeffer*[72]. Their results are summarized in *(162)*.

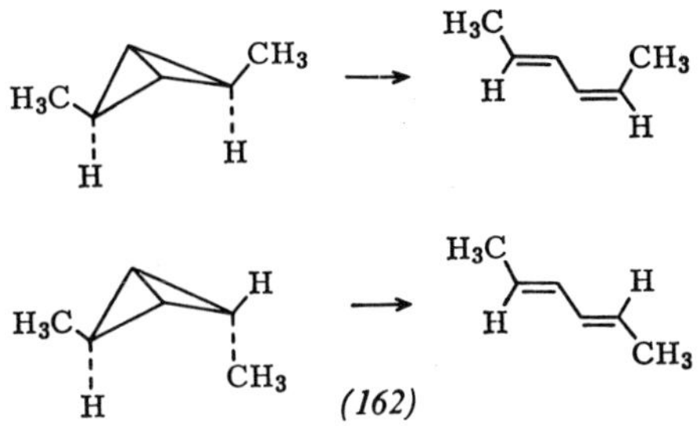

(162)

Remarkably, there exists in principle an alternative pathway relating bicyclobutane and butadiene, with precisely opposite stereochemical consequences. This route [*(163)*→*(164)*→*(165)*] entails the intermediacy of a cyclobutene, which then undergoes conrotatory cleavage to the product butadiene. Thermodynamically, the

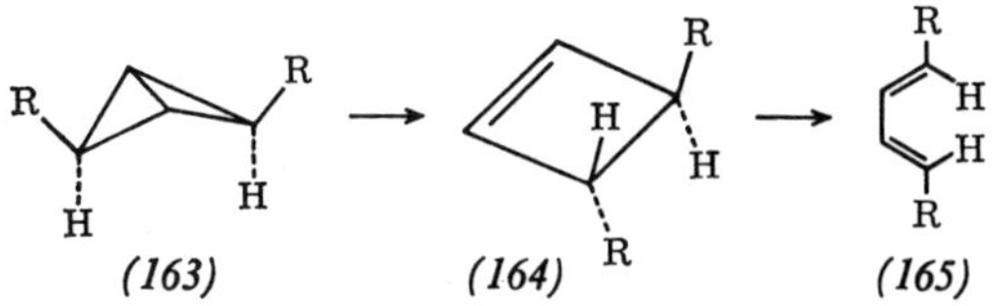

(163) (164) (165)

sequence is feasible[73]. In formal terms, the change may be described as involving successive symmetry-allowed [$_{\sigma}2_s+_{\sigma}2_a$] and [$_{\sigma}2_s+_{\pi}2_a$] reactions. The possibility that special substitution patterns — perhaps operating through attendant simple steric effects — may bring this path to light should not be dismissed.

---

[72] *G. L. Closs* and *P. E. Pfeffer*, J. Amer. chem. Soc. *90*, 2452 (1968).
[73] *R. B. Turner*, *P. Goebel*, *W. von E. Doering*, and *J. F. Coburn*, *Jr.*, Tetrahedron Letters *1965*, 997.

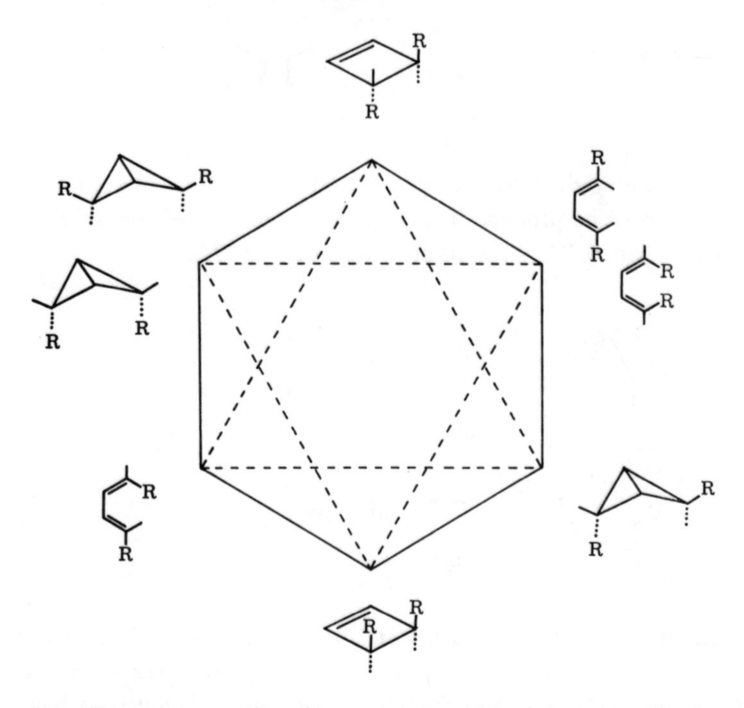

Figure 26. Symmetry-allowed processes relating bicyclobutanes, cyclobutenes, and butadienes. — Allowed ground state paths. - - - Allowed excited state paths.

The reader may find it illuminating to verify for himself the relationships displayed in Figure 26, which presents the stereochemical pattern for all possible symmetry-allowed processes relating bicyclobutanes, cyclobutenes, and butadienes.

The number of known Diels-Alder reactions — the $[_{\pi}4_{s}+{}_{\pi}2_{s}]$ cycloadditions — is legion[74,75]. Though a diradical mechanism has been advanced on numerous occasions[76], the overwhelming body of experimental facts is consistent only with a concerted mechanism[77].

---

[74] *R. Huisgen, R. Grashey*, and *J. Sauer* in *S. Patai:* The Chemistry of Alkenes. Interscience, New York 1964, p. 739.
[75] *A. Wassermann:* The Diels-Alder Reaction. Elsevier, Amsterdam 1965.
[76] Most recently by *S. W. Benson*, J. chem. Physics *46*, 4920 (1967).
[77] *R. B. Woodward* and *T. J. Katz*, Tetrahedron *5*, 70 (1959).

There is a growing body of photochemical Diels-Alder reactions[78], but in most instances there is no evidence which provides information as to whether the reactions are concerted or not. However, it is very probable that concerted symme-

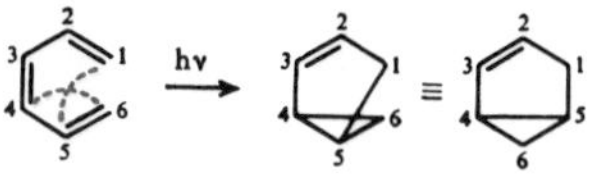

(166)

try-allowed excited-state $[_{\pi}4+_{\pi}2]$ processes are involved in the ubiquitous conversion of *cis*-hexatrienes *(166)* to bicyclo[3.1.0]hexenes[79]. If concerted, these reactions must be either $[_{\pi}4_s+_{\pi}2_a]$ or $[_{\pi}4_a+_{\pi}2_s]$ processes. Excluding the formation of a *trans* fused bicyclohexene, the stereochemical consequences of the allowed processes are seen from formulae *(167)*–*(171)*.

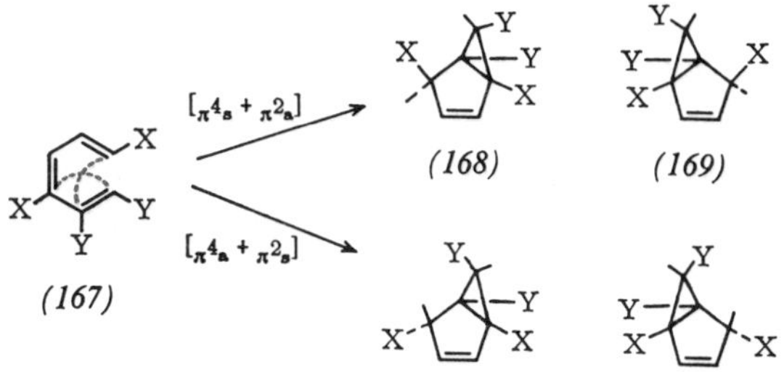

$[_{\pi}4_s + _{\pi}2_a]$

$[_{\pi}4_a + _{\pi}2_s]$

(167) *(168)* *(169)* *(170)* *(171)*

No reaction involving a substance sufficiently extensively labeled as to provide a test of these conclusions has been studied. In vitamin D_2 *(172)*, the top and bottom faces of the triene system are no longer equivalent, and the processes which lead in the general cases to enantiomers must in this instance afford different structures. In fact, irradiation of vitamin D_2 leads to two bicyclohexenes—suprasterol I

[78] For example, *G. S. Hammond*, *N. J. Turro*, and *R. S. H. Liu*, J. org. Chem. *28*, 3297 (1963); *G. S. Hammond* and *R. S. H. Liu*, J. Amer. chem. Soc. *85*, 477 (1963); *D. Valentine*, *N. J. Turro*, and *G. S. Hammond*, ibid. *86*, 5202 (1964); *R. S. H. Liu*, *N. J. Turro*, and *G. S. Hammond*, ibid. *87*, 3406 (1965); *G. O. Schenck*, *S.-P. Mannsfeld*, *G. Schomburg*, and *C. H. Krauch*, Z. Naturforsch. *19b*, 18 (1964).

[79] *J. Meinwald* and *P. H. Mazzocchi*, J. Amer. chem. Soc. *88*, 2850 (1966); *J. Meinwald*, *A. Eckell*, and *K. L. Erickson*, ibid. *87*, 3532 (1965); *H. Prinzbach* and *H. Hagemann*, Angew. Chem. *76*, 600 (1964); Angew. Chem. internat. Edit. *3*, 653 (1964); *H. Prinzbach* and *E. Druckrey*, Tetrahedron Letters *1965*, 2959; *K. J. Crowley*, ibid. *1965*, 2863.

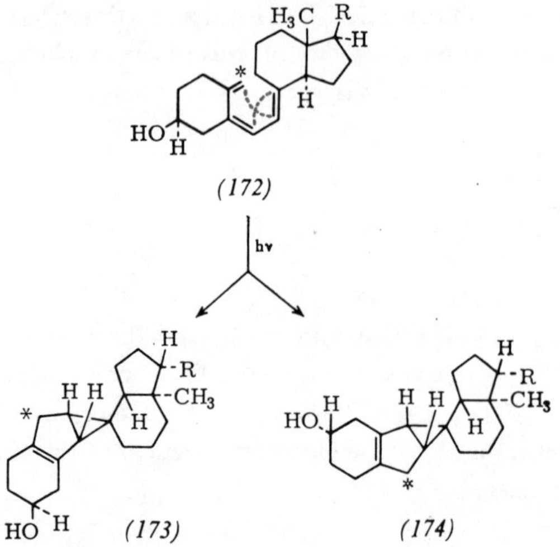

(172)

hν

(173) *(174)*

(173)[80] and suprasterol II *(174)*[81]. These structures are precisely those which must result from concerted symmetry-allowed excited-state $[_{\pi}4_s + _{\pi}2_a]$ processes, but in the absence of further labeling at the starred atoms it is not possible to discern whether the $_{\pi}4$ system participates in the reactions in a suprafacial manner.

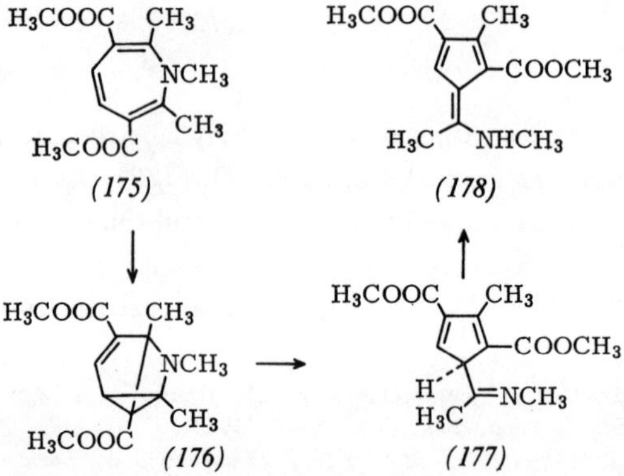

(175) *(178)*

(176) *(177)*

[80] *W. G. Dauben*, personal communication; *W. G. Dauben, I. Bell, T. W. Hutton, G. F. Laws, A. Rheiner*, and *H. Urscheler*, J. Amer. chem. Soc. *80*, 4116 (1958).

[81] *W. G. Dauben* and *P. Baumann*, Tetrahedron Letters *1961*, 565; *C. P. Saunderson* and *D. C. Hodgkin*, ibid. *1961*, 573.

A proposed mechanism for the observed transformation of the azepine *(175)* into the fulvene *(178)* invokes the symmetry-allowed ground-state $[_\pi 4_a + _\pi 2_a]$ reaction *(175)*→*(176)*[82].

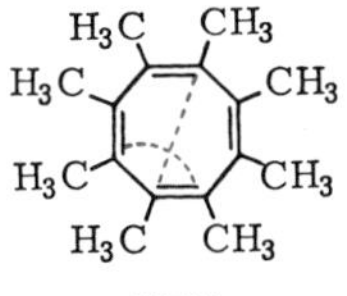

(179) 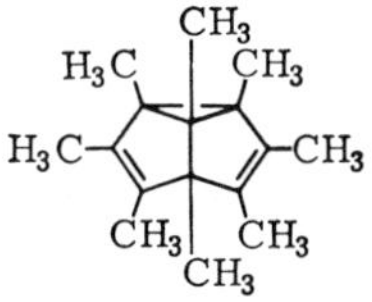 *(180)*

At elevated temperatures, in the presence of base, octamethylcyclooctatetraene *(179)* is smoothly converted to octamethylsemibullvalene *(180)*[83]. It seems possible that the base serves merely to protect the reactant from alternative acid-catalyzed changes, and that the reaction is a simple symmetry-allowed $[_\pi 4_a + _\pi 2_a]$ intramolecular cycloaddition. Further, *Pettit*[84] has made observations which are best interpreted by assuming that dibenzo[*a*,*c*]cyclooctatetraene *(181)* undergoes spontaneous transformation to *(182)*; the special points worthy of note in this case are

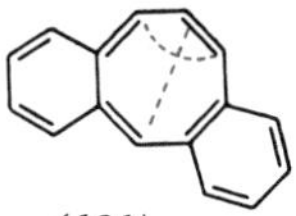

(181) 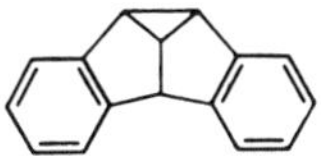 *(182)*

that the $[_\pi 4_a + _\pi 2_a]$ reaction is favored both by concomitant loss of *o*-quinoid character as reaction proceeds, and by deviation in *(181)* from normal cyclooctatetraene geometry, induced by the fused six-membered rings. The difficulty of the $-[_\pi 4_a + _\pi 2_a]$ process [84a] in the absence of favorable steric constraints is shown by

---

[82] *R. F. Childs*, *R. Grigg*, and *A. W. Johnson*, J. chem. Soc. *C*, *1967*, 201.
[83] *R. Criegee* and *R. Askani*, Angew. Chem. *80*, 531 (1968); Angew. Chem. internat. Edit. *7*, 537 (1968).
[84] *G. F. Emerson*, *L. Watts*, and *R. Pettit*, J. Amer. chem. Soc. *87*, 131 (1965); *W. Merk* and *R. Pettit*, ibid. *89*, 4787 (1967).
[84a] Here and in the sequel we use the device of placing a negative sign before the bracket whenever a reaction is characterized in terms of its products.

the fact that the ketone *(183)* fails to undergo cycloreversion to butadiene and *(185)* even at 400°C; the corresponding $-[_{\pi}4_s+_{\pi}2_s]$ reaction *(184)*→*(185)* takes place readily[85].

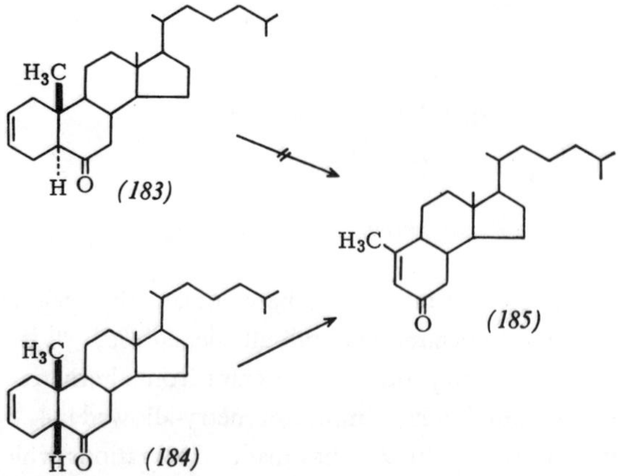

[4+4] Cycloadditions are well known[86]. As yet, they have been encountered only in photochemical reactions, and no detailed stereochemical information is available.

[8+2] Cycloadditions have been relatively rarely observed. Heptafulvene *(186)* readily combines with dimethyl acetylenedicarboxylate[87] to give *(187)*, whereas

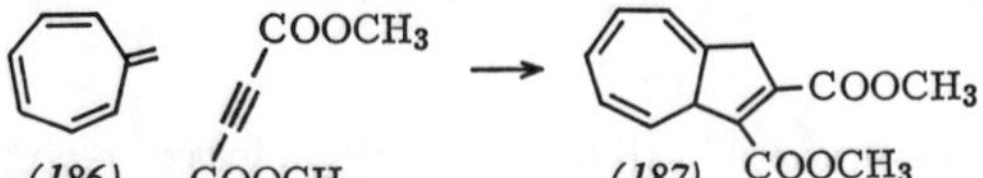

fulvene *(188)* does not react in a similar sense (note that the $[_{\pi}6_s+_{\pi}2_s]$ combina-

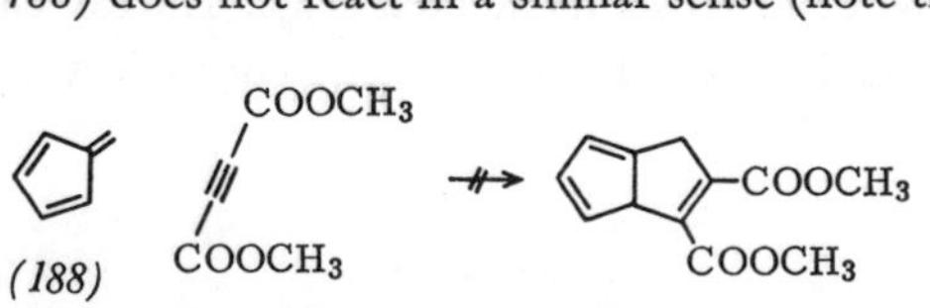

---

[85] *H. R. Nace*, personal communication.

[86] *L. A. Paquette* and *G. Slomp*, J. Amer. chem. Soc. *85*, 765 (1963); *P. de Mayo* and *R. W. Yip*, Proc. chem. Soc. *1964*, 84; *D. E. Applequist* and *R. Searle*, J. Amer. chem. Soc. *86*, 1389 (1964); *J. S. Bradshaw* and *G. S. Hammond*, ibid. *85*, 3953 (1963); *K. Kraft* and *G. Koltzenburg*, Tetrahedron Letters *1967*, 4357.

[87] *W. von E. Doering* and *D. W. Wiley*, Tetrahedron *11*, 183 (1960).

tion is symmetry-forbidden). Similar reactions of calicenes have been observed[88]. Hexaphenylpentalene *(189)* also combines with the acetylenic ester, producing an azulene by way of an adduct *(190)*, as yet not isolated, which is the product of an [8+2] reaction[89].

(189) *(190)*

Two unusual [8+2] cycloadditions, discovered by *Boekelheide*[90], are portrayed in *(191)* and *(192)*.

(191)

(192)

[6+4] Cycloadditions were unknown until our enunciation of the principle of orbital symmetry conservation stimulated the search for them. The cases *(193)*-*(198)* are now known.

[88] *H. Prinzbach, D. Seip*, and *G. Englert*, Liebigs Ann. Chem. *698*, 57 (1966); *H. Prinzbach, D. Seip, L. Knothe*, and *W. Faisst*, ibid. *698*, 34 (1966).
[89] *E. Le Goff*, J. Amer. chem. Soc. *84*, 3975 (1962).
[90] *A. Galbraith, T. Small, R. A. Barnes*, and *V. Boekelheide*, J. Amer. chem. Soc. *83*, 453 (1961); *V. Boekelheide* and *N. A. Fedoruk*, Proc. nat. Acad. Sci. *55*, 1385 (1966).

(193)[91]

(194)[91]

(195)[92]

(196)[92,93]

(197)[94]

(198)[95]

R = $-COOC_2H_5$

[6+6] Cycloadditions of presumed *p*-xylylene intermediates are known[96], but no information is available about the detailed mechanism of the reactions. Irradiation of tropone in acidic solution gives a symmetrical dimer *(199)*[97], which is the product to be expected from a concerted symmetry-allowed excited-state $[_{\pi}6_{s}+_{\pi}6_{s}]$ combination; the unusual experimental conditions, and the fact that other photo-induced dimerizations of tropone may proceed in a non-concerted fashion[98] suggest caution in adopting the conclusion that the concerted process is in fact involved in the formation of *(199)*.

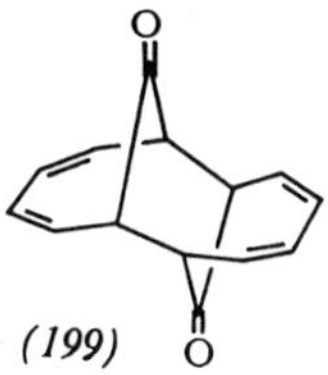

(199)

Heptafulvalene *(200)* combines with tetracyanoethylene to give the adduct *(201)*, whose structure has been firmly established by X-ray crystallographic methods[99].

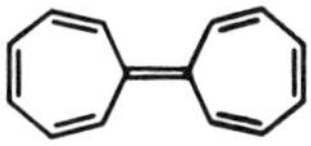

(200)

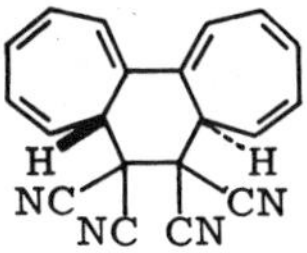

(201)

The adduct is the expected product of the symmetry-allowed ground-state $[_{\pi}14_{a}+_{\pi}2_{s}]$ process; it may be noted that the twisted shape of the heptafulvalene molecule sets the stage ideally for *antara* addition to the 14-electron system.

---

[91] *K. Houk*, Dissertation, Harvard (1968).
[92] *R. C. Cookson, B. V. Drake, J. Hudec*, and *A. Morrison*, Chem. Commun. *1966*, 15; *S. Itô, Y. Fujise, T. Okuda*, and *Y. Inoue*, Bull. chem. Soc. Japan *39*, 135 (1966).
[93] *T. Nozoe, T. Mukai, K. Takase*, and *T. Takase*, Proc. Japan Acad. *28*, 477 (1952).
[94] *S. Itô, Y. Fujise*, and *M. C. Woods*, Tetrahedron Letters *1967*, 1059.
[95] *L. A. Paquette* and *J. H. Barrett*, J. Amer. chem. Soc. *88*, 2590 (1966).
[96] *H. E. Winberg, F. S. Fawcett, W. E. Mochel*, and *C. W. Theobald*, J. Amer. chem. Soc. *82*, 1428 (1960); *D. J. Cram, C. S. Montgomery*, and *G. R. Knox*, ibid. *88*, 515 (1966); *D. J. Cram, C. K. Dalton*, and *G. R. Knox*, ibid. *85*, 1088 (1963); *D. T. Longone* and *F.-P. Boettcher*, ibid. *85*, 3436 (1963).
[97] *T. Mukai, T. Tezuka*, and *Y. Akasaki*, J. Amer. chem. Soc. *88*, 5025 (1966).
[98] *A. S. Kende* and *J. E. Lancaster*, J. Amer. chem. Soc. *89*, 5283 (1967).
[99] *W. von E. Doering*, personal communication.

Cycloadditions involving ionic components are as yet not common. Ionic [4+2] cycloadditions may be realized in a number of different ways [*(202)*–*(204)*].

Class A

(202)

Class B

(203)

Class C

(204)

We may assign to class A the cycloaddition reactions of cyclopropanones *(205)*. It is not unlikely that cyclopropanones are in equilibrium with an isomeric dipolar species. The latter can act as a 2 π-electron system, and, predictably, combines with

(205)

dienes in $[_{\pi}4_s + _{\pi}2_s]$ processes — for example, with furans to give adducts of the type *(206)*[100].

(206) *(207)* *(208)*

Most recently, the direct combination of the 2-methylallyl cation *(207)* with cyclopentadiene and with cyclohexadiene, to give the bicyclic cations *(208)*, Z = $-CH_2-$ or $-CH_2-CH_2-$, has been demonstrated[101].

---

[100] *R. C. Cookson, M. J. Nye*, and *G. Subrahmanyam*, J. chem. Soc. *C*, *1967*, 473; *1965*, 2009; *A. W. Fort*, J. Amer. chem. Soc. *84*, 2620, 2625, 4979 (1964); *N. J. Turro* and *W. B. Hammond*, ibid. *87*, 3258 (1965); *88*, 3672 (1966); *W. B. Hammond* and *N. J. Turro*, ibid. *88*, 7880 (1966).
[101] *H. M. R. Hoffmann, D. R. Joy*, and *A. K. Suter*, J. chem. Soc. *B*, *1968*, 57.

A remarkable cycloaddition of class B is that which occurs in the formation of the pipitzols *(210)* from perezone *(209)*[102].

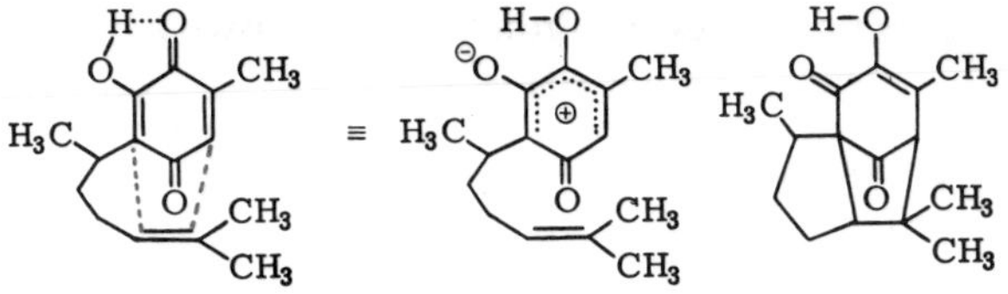

(209) *(210)*

As yet, no simple instances of class C ionic [4+2] cycloadditions have been discovered. However, a very important and numerous group of reactions — the 1,3-dipolar additions[103] — may be placed in this class, since the 1,3-dipolar substances which participate in cycloaddition reactions, although formally neutral molecules, generally function as three-orbital 4 π-electron species. *Huisgen*, whose own magnificent researches have transformed 1,3-dipolar addition from an obscure phenomenon, exemplified only by a few curiosities, into a major reaction-type, has presented excellent surveys of the field[103], and has analyzed the process in terms of orbital symmetry conservation[104]. Consequently, it is only necessary here to summarize briefly the major fundamental factors, and to allude to a few special points. The great majority of known participants in 1,3-dipolar addition reactions are isoelectronic either with ozone *(211)* or with nitrous oxide *(212)*, and unambiguously contain a three-orbital system occupied by four π electrons.

Another class of reactants is represented by labile molecules which undergo ready transformation into species containing the requisite 4-electron system, *e.g.* the car-

(211) *(212)*

---

[102] *F. Walls, J. Padilla, P. Joseph-Nathan, F. Giral,* and *J. Romo,* Tetrahedron Letters *1965,* 1577.

[103] *R. Huisgen,* Angew. Chem. *75,* 604 (1963); Angew. Chem. internat. Edit. *2,* 565 (1963); *R. Huisgen, R. Grashey,* and *J. Sauer* in *S. Patai:* The Chemistry of Alkenes. Interscience, New York 1964, p. 739; *R. Huisgen,* Helv. chim. Acta *50,* 2421 (1967).

[104] *A. Eckell, R. Huisgen, R. Sustmann, G. Wallbillich, D. Grashey,* and *E. Spindler,* Chem. Ber. *100,* 2192 (1967).

bonyl ylides *(214)* and the azomethine ylides *(216)* are produced by electrocyclic transformation of ethylene oxides *(213)* and ethylene imines *(215)*, substituted in such wise as to facilitate ring cleavage; the symmetry-determined conrotatory course of these antecedent processes has already been discussed above (*cf.* Section 5).

(213) *(214)*

(215) *(216)*

The situation in respect to the possibility of the participation, in concerted cycloaddition reactions, of fugitive species such as vinylcarbene and its isoelectronic analogues, presents interesting ambiguities. Such molecules may exist in three forms, differing in the number of π electrons occupying the allyl orbital. The first, *(217)*,

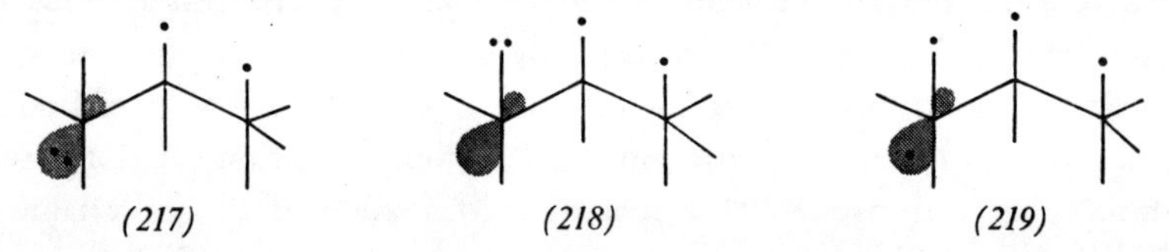

(217) *(218)* *(219)*

containing two electrons in the allyl system, may be expected to combine with dienes in a [2+4] process, but independent reaction as a simple substituted carbene could supervene. The second, *(218)*, with four electrons in the allyl system, should undergo normal concerted combination with 2 π-electron molecules, while the third, *(219)*, is an analogue of triplet carbene. These systems have not been extensively studied; it may be expected that their electronic structures, and consequently the courses of their reactions will vary as the backbone atoms, and attached substituents, are varied.

Brief mention may be made of the possibility of analogues of 1,3-dipolar additions, in which larger numbers of electrons are incorporated in the π system of the dipolar component. As speculative examples, the *cis* form *(220)* of a dimeric nitroso compound, the α-ketonitrones *(221)*, and the α-nitrosonitrones *(222)*, all contain 6 π-electron systems, and concerted combination with dienes is symmetry-allowed. A special word of caution should be added in relation to 1,3-dipolar additions.

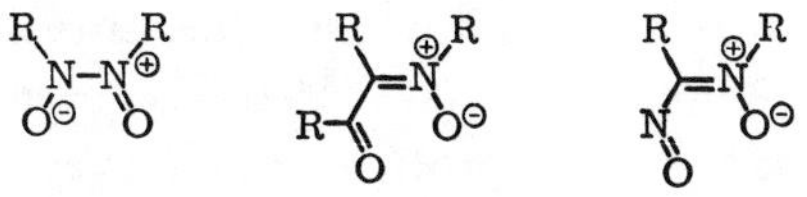

(220) *(221)* *(222)*

There may well be instances of such reactions in which the complementary polar character of the reactants is sufficiently extreme as to favor two-step combination, proceeding through a relatively stable dipolar intermediate.

6.2. [2+2] Cycloadditions in the Photochemistry of the Cyclohexadienones and Cyclohexenones

Following the pioneering work of *Barton*, *Jeger*, and their co-workers, the beautiful complexity of the photochemistry of cyclohexadienones and cyclohexenones has been most ably explored by *Chapman*, *Schaffner*, *Schuster*, *Zimmerman*, and their collaborators[105]. The primary isomerization processes which have been noted are of three types *(223)*—*(225)*.

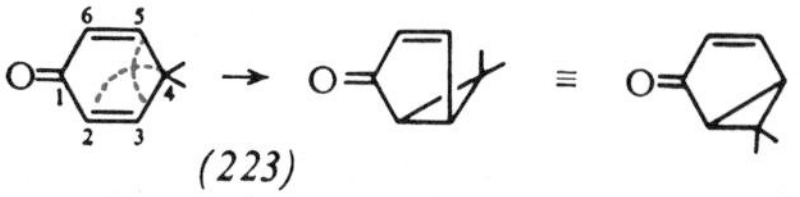

(223) Class A[106]

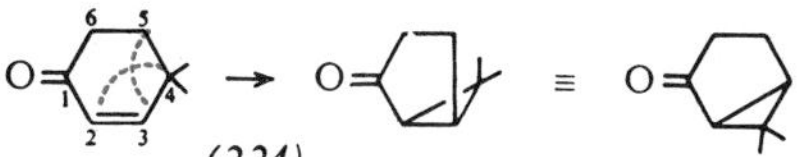

(224) Class B[107]

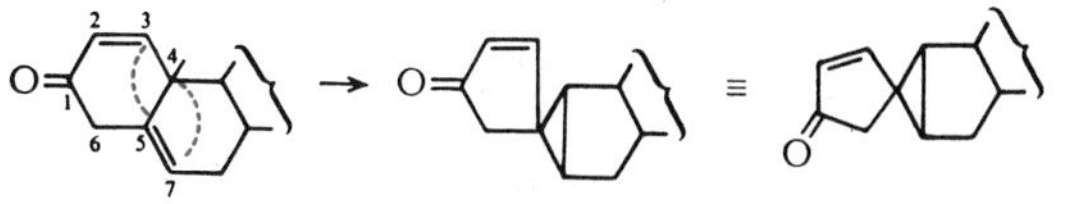

(225) Class C[108]

---

[105] Two recent reviews serve as excellent guides to the elegant structural work accomplished. (a) *K. Schaffner*, Advances in Photochemistry *4*, 81 (1966); (b) *P. J. Kropp*, Organic Photochemistry *1*, 1 (1967).

[106] *D. H. R. Barton*, *J. McGhie*, and *R. Rosenberger*, J. chem. Soc. *1961*, 1215; *D. H. R. Barton*, *P. de Mayo*, and *M. Shafiq*, ibid. *1958*, 140, 3314; *D. Arigoni*, *H. Bosshard*, *H. Bruderer*, *G. Büchi*, *O. Jeger*, and *L. J. Krebaum*, Helv. chim. Acta *40*, 1732 (1957) and other papers referred to in ref. [105].

[107] *W. W. Kwie*, *B. A. Shoulders*, and *P. D. Gardner*, J. Amer. chem. Soc. *84*, 2268 (1962); *O. L. Chapman*, *T. A. Rettig*, *A. A. Griswold*, *A. I. Dutton*, and *P. Fitton*, Tetrahedron Letters *1963*, 2049; *B. Nann*, *D. Gravel*, *R. Schorta*, *H. Wehrli*, *K. Schaffner*, and *O. Jeger*, Helv. chim. Acta *46*, 2473 (1963) and other papers referred to in ref. [105].

[108] *B. Nann*, *D. Gravel*, *R. Schorta*, *H. Wehrli*, *K. Schaffner*, and *O. Jeger*, *Helv. chim. Acta 46* 2473 (1963); *B. Nann*, *H. Wehrli*, *K. Schaffner*, and *O. Jeger*, ibid. *48*, 1680 (1965).

It seems probable that all, and has been established that some, of these cases involve reactions proceeding through n,π* triplet states. It has also been clearly established that the quantum efficiency of the processes varies over a very wide range[109]. Detailed mechanisms have been proposed for these reactions[110]; the most elaborate and ingenious of these mechanisms is that suggested by *Zimmerman*[111].

We would like to examine the stereochemical constraints imposed by conservation of orbital symmetry on these transformations. These constraints are operative only if the reactions are concerted. We are aware of the elegant physical measurements which have been interpreted as supporting the proposed non-concerted mechanisms. But we regard it as worth while to stipulate the stereochemical imperatives associated with concerted paths from the starting materials to the observed products.

Formally, all of the reactions are $[_\sigma 2 + _\pi 2]$ cycloadditions. If they are concerted changes within excited states, such cycloadditions must be $[_\sigma 2_s + _\pi 2_s]$ or $[_\sigma 2_a + _\pi 2_a]$ reactions.

In classes A and B suprafacial participation of double bond 2,3 is a stereochemical impossibility, since it leads to *trans* fused three- and five-membered rings. Consequently, these reactions must follow the $[_\sigma 2_a + _\pi 2_a]$ course, which requires antarafacial addition at double bond 2,3 , and inversion at C-4. In class C the special constraint present in classes A and B is absent. Both $[_\sigma 2_a + _\pi 2_a]$ and $[_\sigma 2_s + _\pi 2_s]$ processes are feasible.

Antara addition at the double bond and inversion at the migrating saturated carbon atom is of course precisely what occurs in the classic conversion of santonin *(226)* into lumisantonin *(227)*[112]. Unfortunately, in this case the alternative symmetry-forbidden change, taking place with retention at C-4, would lead to a steri-

---

[109] *H. E. Zimmerman, R. G. Lewis, J. J. McCullough, A. Padwa, S. Staley*, and *M. Semmelhack*, J. Amer. chem. Soc. *88*, 159 (1966); *O. L. Chapman, J. B. Sieja*, and *W. J. Welstead, Jr.*, ibid. *88*, 161 (1966).

[110] *H. E. Zimmerman*, Advances in Photochemistry *1*, 183 (1963); *O. L. Chapman*, ibid. *1*, 323 (1963).

[111] *H. E. Zimmerman*, 17th National Organic Symposium of the American Chemical Society, Bloomington, Indiana, 1960, Abstracts, p. 31; *H. E. Zimmerman* and *D. I. Schuster*, J. Amer. chem. Soc. *83*, 4486 (1961); *84*, 4527 (1962); *H. E. Zimmerman*, Tetrahedron *19*, Supplement 2, 393 (1963).

[112] *D. H. R. Barton, P. de Mayo*, and *M. Shafiq*, J. chem. Soc. *1958*, 140; *D. Arigoni, M. Bosshard, H. Bruderer, G. Büchi, O. Jeger*, and *L. J. Krebaum*, Helv. chim. Acta *40*, 1732 (1957); *W. Cocker, K. Crowley, J. T. Edward, T. B. H. McMurry*, and *E. R. Stuart*, J. chem. Soc. *1957*, 3416; *D. H. R. Barton* and *P. T. Gilham*, J. chem. Soc. *1960*, 4596.

cally unlikely *trans* ring fusion; consequently, it might be argued that the observed path, though symmetry-allowed, would in any event be imposed by the geometry of the system.

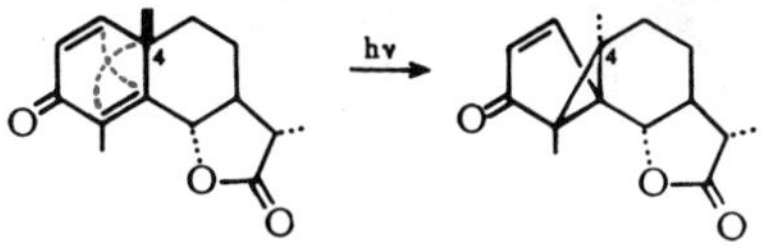

hv

(226) (227)

The simplest labeling of the parent cyclohexadienone which would endow it with a center of chirality would also allow in principle a resolution of this question. Consider the consequences of retention or inversion at C-4 on such a chiral cyclohexadienone *(228)*[113]. Inversion or retention at C-4 can each be performed in two distinct ways, depending on whether the addition at C-3 is on top or bottom. If A and

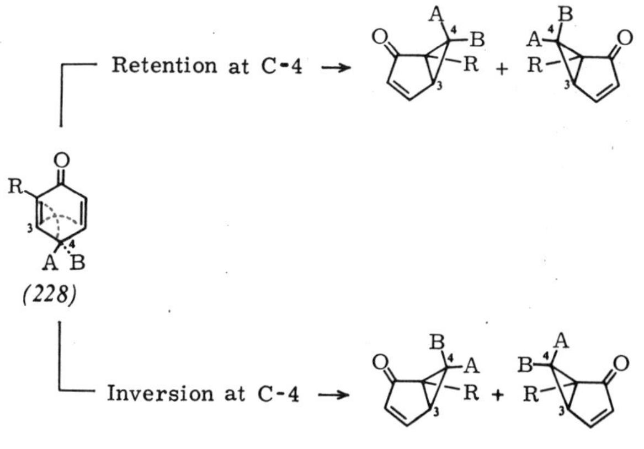

Retention at C-4 →

(228)

Inversion at C-4 →

B are different substituents these pathways become distinct, each with its own steric factors. It is thus likely that even a stereospecific process will yield two products in differing proportions. The products of inversion are diastereomeric, and enantiomeric to the products of retention. The actual course followed in the rearrangement of a simple chiral cyclohexadienone has not yet been determined, but is being

[113] Only the consequences of an addition on the double bond next to R are sketched. Of course the addition on the opposite side is feasible, and will double the number of possible products.

studied by *Schuster*[114] who has independently recognized that the matter is one of great interest.

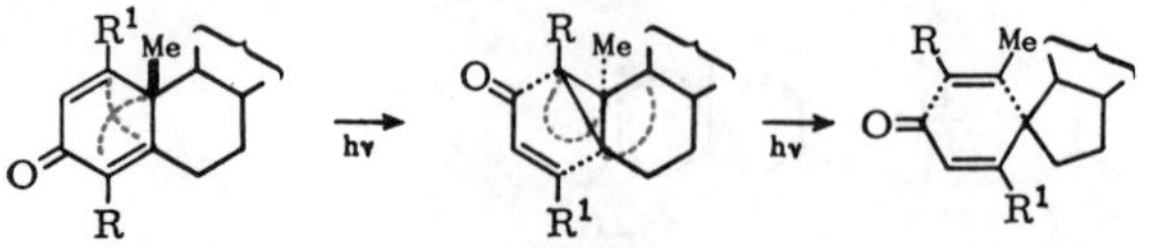

(229) (230) (231)

The versatile field of steroid derivatives appears already to have provided verifications of our predictions. The remarkably rich photochemistry of 1-dehydrotestosterone acetate and its methyl derivatives[115] begins with the now familiar con-

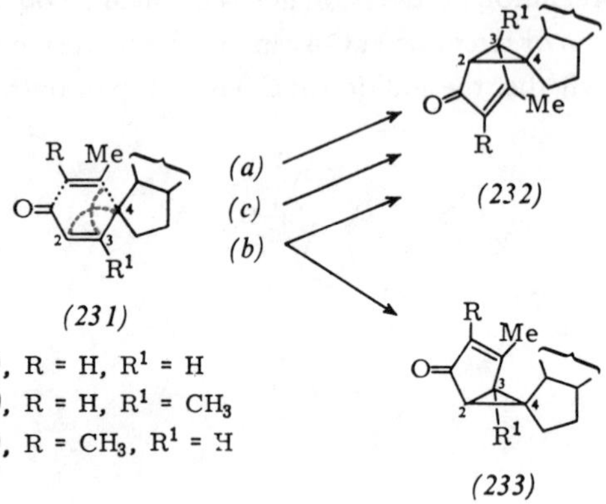

(231)

(a), R = H, R^1 = H
(b), R = H, R^1 = CH_3
(c), R = CH_3, R^1 = H

(232)

(233)

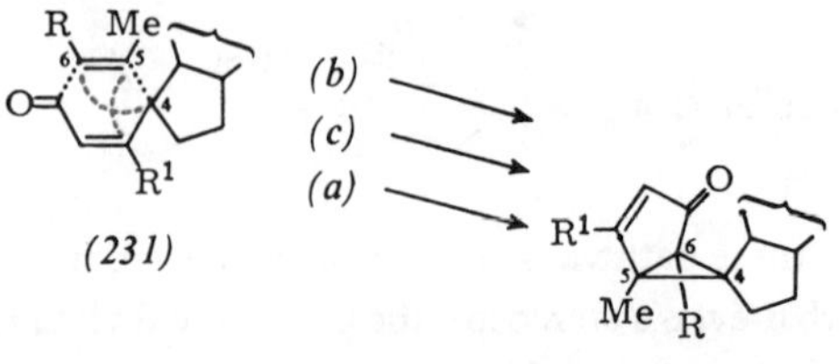

(231)

(234)

---

[114] *D. I. Schuster*, personal communication.
[115] *H. Dutler, M. Bosshard*, and *O. Jeger*, Helv. chim. Acta *40*, 494 (1957); *K. Weinberg, E. C. Utzinger, D. Arigoni*, and *O. Jeger*, ibid. *43*, 236 (1960); *H. Dutler, C. Ganter, H. Ryf, E. C. Utzinger, K. Weinberg, K. Schaffner, D. Arigoni*, and *O. Jeger*, ibid. *45*, 2346 (1962); *C. Ganter, F. Greuter, D. Kägi, K. Schaffner*, and *O. Jeger*, ibid. *47*, 627 (1964); *F. Frei, C. Ganter, D. Kägi, K. Kocsis, M. Miljković, A. Siewinski, R. Wenger, K. Schaffner*, and *O. Jeger*, ibid. *49*, 1049 (1966).

version of a 2,5-cyclohexadienone *(229)* into a bicyclo[3.1.0]hexenone *(230)*, and thence to a new cyclohexadienone *(231)* of assigned stereochemistry. On further photolysis *(231)* furnishes two or three isomers, depending on the pattern of substitution. The ingeniously assigned[115] structures of the products are shown in formulae *(232)*—*(234)*.

The molecules *(232)*, *(233)*, and *(234)* are all products of $[_{\pi}2_a+_{\sigma}2_a]$ processes, precisely as predicted from orbital symmetry considerations for concerted changes; it is especially to be noted that inversion must occur at C-4 during each of the photoisomerizations. For the substitution patterns *(a)* and *(c)* only one product *(232)* of the two which might result from symmetry-allowed changes involving the 2,3 double bond is observed. Similarly, only *(234)* is produced from *(a)* and *(c)* when the 5,6 double bond is implicated. In the case of *(b)*, three of the four possible products of symmetry-allowed processes have been isolated. Not one of the four products which might result from symmetry-forbidden or non-stereospecific processes has been found.

Unfortunately, the Zimmerman mechanism[111] for reactions of class A, illuminated with the insight afforded by orbital symmetry considerations, leads to precisely the same products which must be obtained in a concerted reaction. Formulae *(235)*—*(238)* show the basic mechanism suggested by *Zimmerman*[111].

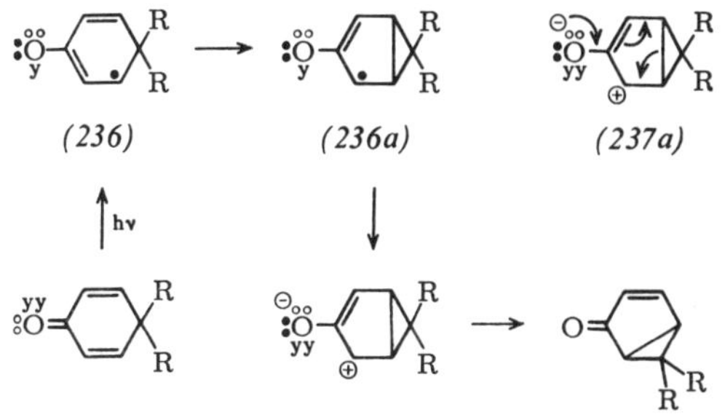

(236) *(236a)* *(237a)*

hν

(235) *(237)* *(238)*

The final step, following bond reorganization [*(236)*→*(236a)*] (symmetry-allowed!) and demotion [*(236a)*→*(237)*] is formally a ground-state cyclopropylcarbinyl rearrangement; note that reversion of *(237)* to the original cyclohexadienone *(235)* [*cf.* arrows in *(237a)*] is symmetry-forbidden! There are actually two possibilities for the last step of this sequence. They are best described as [1,2n] sigmatropic shifts (*cf.* Section 7, below). The first process consists of two sequential [1,2] shifts [*(239)*→*(240)*→*(241)*≡*(243)*], while the second is a [1,4] rearrangement [*(242)*→*(241)*≡*(243)*]. Since [1,2] sigmatropic shifts must proceed with *retention*, but [1,4] shifts, if forced to be suprafacial, must proceed with *inversion* at the migrating carbon, it follows that the consequences of the two alternatives are

stereochemically indistinguishable. Thus, for reactions of class A, either a totally concerted mechanism or a stepwise one involving concerted elements leads to the same stereochemical result. While the physical evidence adduced for the chronology proposed by *Zimmerman* is very impressive, we do not regard it as conclusive: either mechanism remains a possibility.

The photochemical rearrangement of cyclohexenones (the class B reactions) is a highly stereospecific process, and very probably follows a concerted course. Thus, *Chapman* found[116] retention of chirality in the reactions *(244)* and *(245)*.

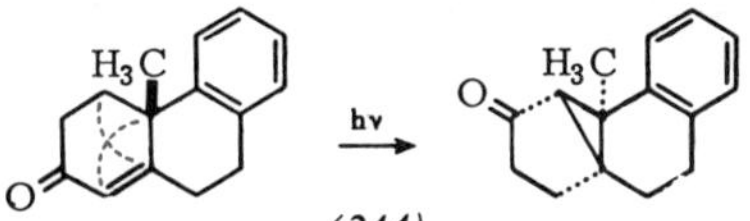

hν

(244)

hν

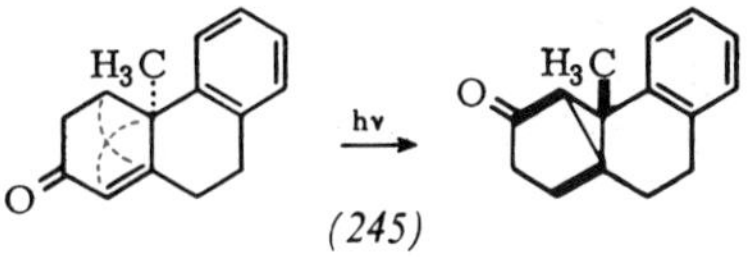

(245)

Belluš and *Schaffner*[117] studied the reaction with a very subtle probe; the isomerization of *(246)* proceeds with retention at C-1 and inversion at C-10. The inversion at C-10 is forced by the steric situation, but the retention at C-1 is not, and is clearly inconsistent with a radical fragmentation of the C(1)—C(10) bond.

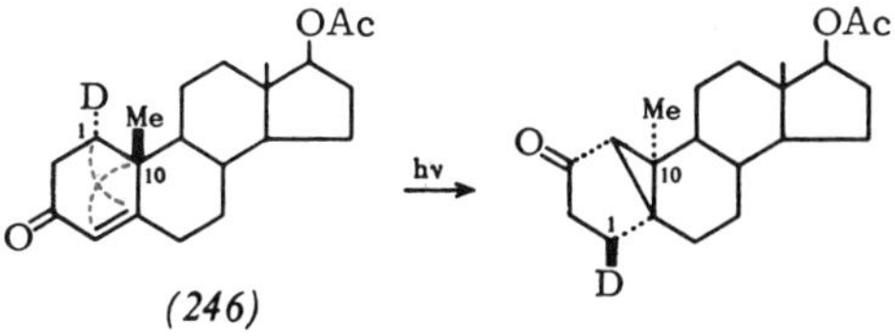

hν

(246)

We turn now to reactions of class C. In this case there are no steric restraints operative against either the $[_{\pi}2_s+_{\sigma}2_s]$ or the $[_{\pi}2_a+_{\sigma}2_a]$ concerted processes. In the beautiful case *(247)* that has been studied[108] just the products predicted for the two allowed cycloadditions are found. But once again avoidance of *trans* fused rings would lead to the very same stereochemical consequences for stepwise processes.

---

[116] *O. L. Chapman, J. B. Sieja,* and *W. J. Welstead, Jr.*, J. Amer. chem. Soc. *88*, 161 (1966).
[117] *D. Belluš, D. R. Kearns,* and *K. Schaffner*, Helv. chim. Acta *52*, 971 (1969).

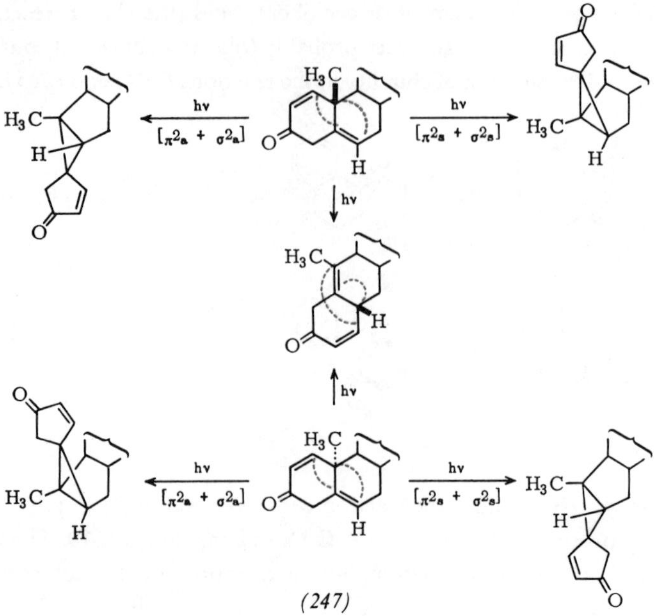

H_3C H hν $[_{\pi}2_a + _{\sigma}2_a]$ O H_3C H hν $[_{\pi}2_s + _{\sigma}2_s]$ H_3C H O

hν

H_3C H O

hν

O H_3C H hν $[_{\pi}2_a + _{\sigma}2_a]$ O H_3C H hν $[_{\pi}2_s + _{\sigma}2_s]$ H_3C H O

(247)

There are two further photochemical transformations of enones which fall into the category of [2+2] cycloadditions. The first of these is an isomerization of a non-conjugated cyclohexenone *(248)*[118].

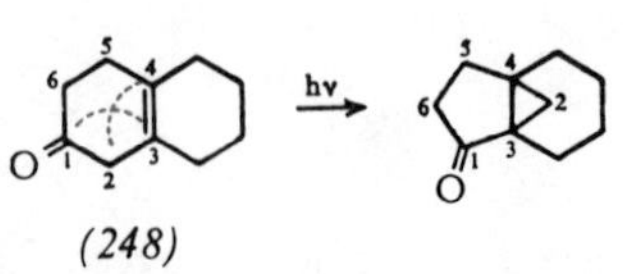

O 1 2 3 4 5 6 hν 5 4 6 2 1 3 O

(248)

Formulae *(249)* show the consequences of the stereochemically feasible reactions *antara* on the double bond with options of retention or inversion at C-2.

A steroidal derivative of this class has been photolyzed[118]. Of the two distinct possible products, only that one is produced whose cyclopropane ring is α-oriented *(250)*. Since labels are lacking at C-2, the primary question of retention or inver-

[118] *J. R. Williams* and *H. Ziffer*, Chem. Commun. *1967*, 194, 469; Tetrahedron *24*, 6725 (1968).

$[_{\pi}2_a + _{\sigma}2_s]$
Retention at C-2

Inversion at C-2
$[_{\pi}2_a + _{\sigma}2_a]$

(249)

(250)

sion is as yet unanswered. We predict that the reaction will be found to proceed with inversion at C-2. Similar circumstances obtain in the case of a further example [*(251)*→*(252)*][119].

There appear to be a number of examples of photochemically induced $[_{\pi}2_a + _{\sigma}2_a]$

(251) (252)

cycloadditions involving simple olefins and dienes. Thus, *Griffin*[120] observed unusual propylene→cyclopropane cyclizations accompanied by group migrations

(253)

(254)

[119] *L. A. Paquette, R. F. Eizember*, and *O. Cox*, J. Amer. chem. Soc. *90*, 5153 (1968).
[120] *G. W. Griffin, J. Covell, R. C. Petterson, R. M. Dodson*, and *G. Klose*, J. Amer. chem. Soc. *87*, 1410 (1965); *H. Kristinsson* and *G. W. Griffin*, ibid. *88*, 378 (1966).

[*(253)*, *(254)*]. These could well be concerted transformations, but definitive stereochemical information is lacking.

An instance of this type of reaction in which the stereochemical course has been es-

(255) 140 : 1

tablished is available in the photolysis of 4,4-diphenylcyclohexenone *(255)*[121]. The predominant product is the result of an antarafacial addition on the double bond and inversion at C-4.

(256)

A probable further example of a [$_{\sigma}2_a+_{\pi}2_a$] reaction has emerged in a recent study of the photolysis of 3,19-dioxo-4-androsten-17β-yl acetate *(256)*[122].

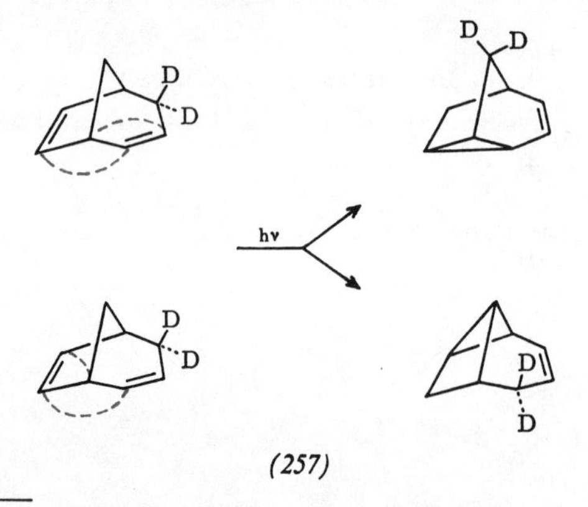

(257)

---

[121] *H. E. Zimmerman* and *K. G. Hancock*, J. Amer. chem. Soc. *90*, 3749 (1968). See also *H. E. Zimmerman* and *R. L. Morse*, ibid. *90*, 954 (1968).

[122] *E. Pfenninger, D. E. Poel, C. Berse, H. Wehrli, K. Schaffner*, and *O. Jeger*, Helv. chim. Acta *51*, 772 (1968).

Still other possible $[_{\sigma}2_{a}+_{\pi}2_{a}]$ changes, constrained to proceed as such, are available in the case *(257)* studied by *Sauers*[123].

The formation of *(259)* on photolysis of cyclooctatriene *(258)* is also constrained to follow the $[_{\pi}2_{a}+_{\sigma}2_{a}]$ course, if concerted; it has been shown that the process occurs without a hydrogen shift[124].

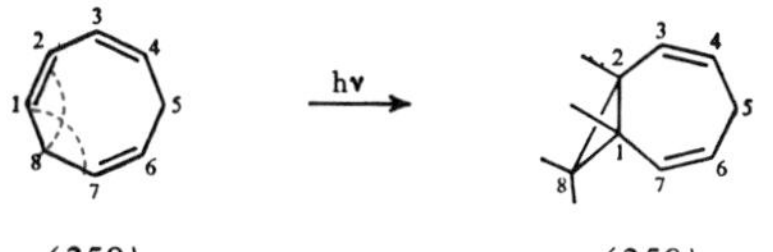

(258) *(259)*

A related case *(260)* has been observed by *Edman*[125]. A concerted cycloaddition

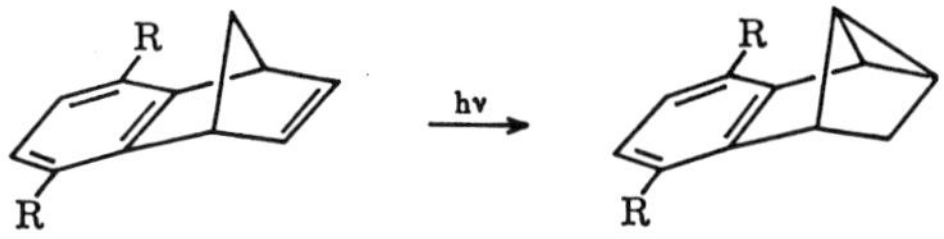

(260)

may involve either the σ bond at 3,7 *(261)* or that at 2,3 *(262)*. The concerted nature of the reaction could be ascertained using a chiral reactant.

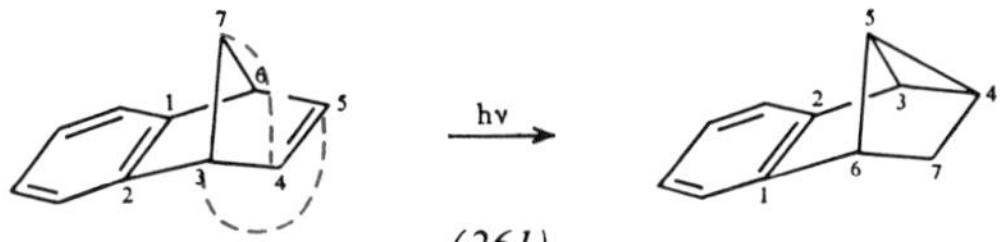

(261)

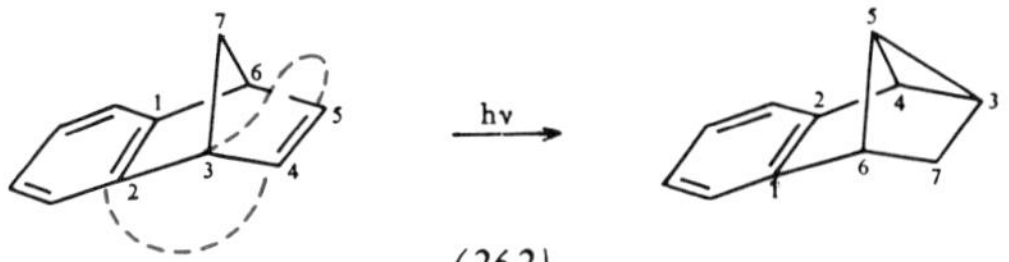

(262)

---

[123] *R. R. Sauers* and *A. Shurpik*, J. org. Chem. *33*, 799 (1968).
[124] *W. R. Roth* and *B. Peltzer*, Liebigs Ann. Chem. *685*, 56 (1965).
[125] *J. R. Edman*, J. Amer. chem. Soc. *88*, 3454 (1966).

Still other reactions of this type are found in the conversion of a dibenzobicyclo[2.2.2]octatriene *(263)* to a dibenzosemibullvalene[126] and of a labeled benzobicyclo[2.2.2]octatriene to a benzosemibullvalene[127].

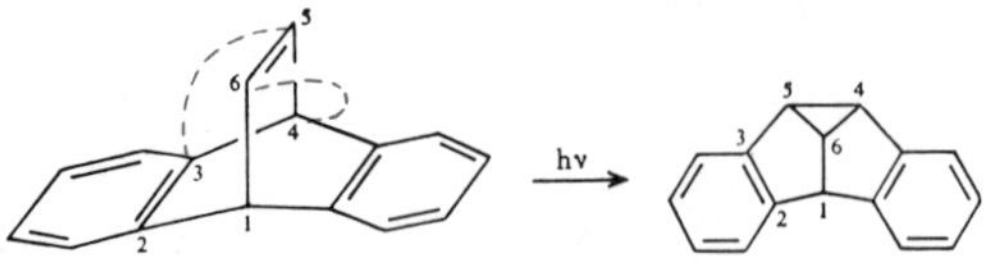

(263)

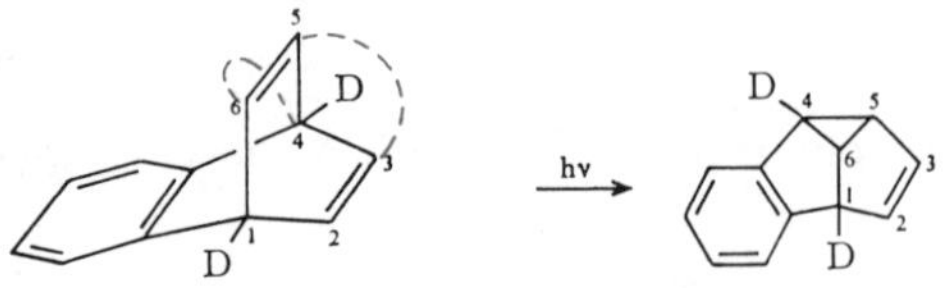

(264)

We emphasize that neither of these cases need be concerted; indeed, there is evidence for a stepwise process in the corresponding reaction of the parent barrelene[128].

We must now examine the question of how a process shown to proceed through an excited triplet state of a molecule could possibly be concerted. We believe that orbital symmetry relationships dictate to a molecule in an excited state a certain set of motions, leading to reaction, which are facile, and another set of motions which are difficult. There is no necessity to reach the excited state of products. The symmetry-allowed motions are initiated in the excited state of the reactant. In their course they are accompanied by a radiationless transition to the ground-state of the product. While the physical rationale of such a transition is still lacking, the occurrence of such a process presents no more difficulty than does any other radiationless transition.

---

[126] *E. Ciganek*, J. Amer. chem. Soc. *88*, 2882 (1966).
[127] *H. E. Zimmerman, R. S. Givens*, and *R. M. Pagni*, J. Amer. chem. Soc. *90*, 4192 (1968). See also *J. P. N. Brewer* and *H. Heaney*, Chem. Commun. *1967*, 811; *P. W. Rabideau, J. B. Hamilton*, and *L. Friedman*, J. Amer. chem. Soc. *90*, 4465 (1968).
[128] *H. E. Zimmerman, R. W. Binkley, R. S. Givens*, and *M. A. Sherwin*, J. Amer. chem. Soc. *89*, 3932 (1967).

6.3. The [2+2+2] Cycloaddition Reaction

Further useful practice in the application of the principle of orbital symmetry conservation may be gained from a detailed consideration of the [2+2+2] concerted cycloaddition (or cycloreversion) reaction; the exercise is made the more valuable through the emergence from it of some new factors, and through the fact that it sets the stage for comparisons, in the sequel, which deepen our understanding of orbital symmetry conservation as a fundamental extension of our knowledge of the phenomenon of chemical bonding.

Let us consider the decomposition of cyclohexane, through a boat-shaped transition state, into three molecules of ethylene. First, we assign six electrons — those

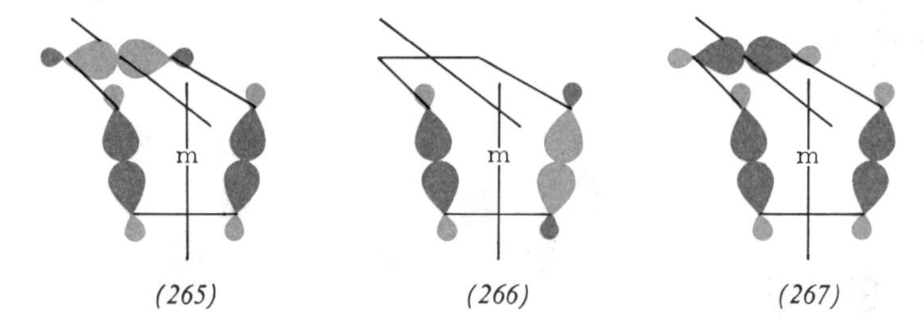

(265) *(266)* *(267)*

involved in the bonds to be broken — in pairs, to the generalized σ orbitals *(265)*, *(266)*, and *(267)* [*cf.* Section 3.1]. Note that the only symmetry element imposed by the geometry of the system is the mirror plane *m*, and that the orbitals shown are, as required, either symmetric or antisymmetric with respect to that element. *It is of especial importance in this case to observe the fundamental principle that unsymmetrical molecular orbitals are not permissible:* thus, the orbital *(268)* is not real, and must be

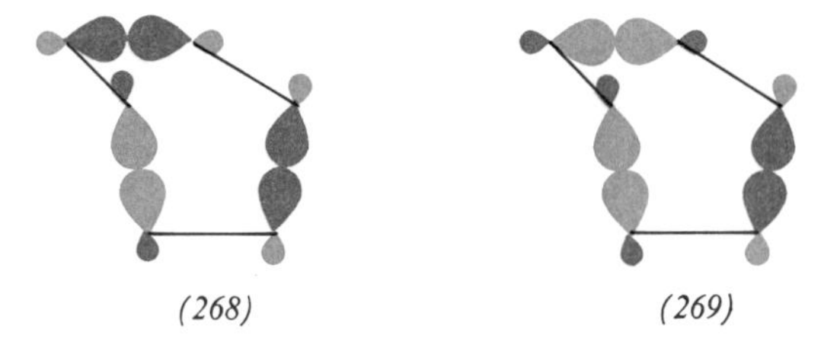

(268) *(269)*

expunged by addition to its equally unreal conjugate *(269)*. It is also worthy of note that the orbitals *(267)* and *(265)* may be replaced by their sum *(270)* and difference *(271)*, with no change in any conclusion which may be discovered from

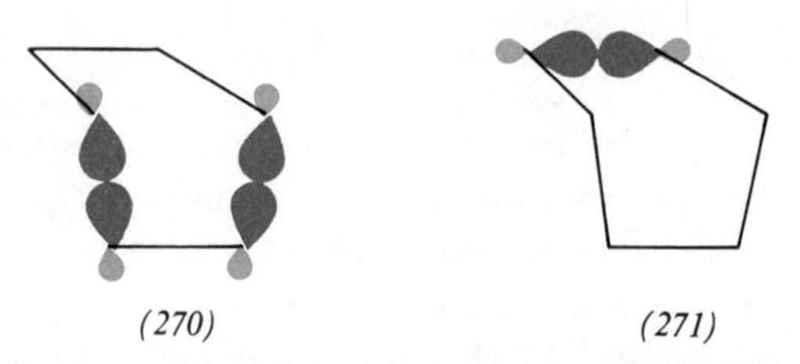

(270) *(271)*

their use; but in general the most obviously informative practice is to make use of those designations which most explicitly reveal the largest extent of delocalization in the generalized molecular orbitals — in this case *(265)* and *(267)*, rather than *(270)* and *(271)*. In any event, no other acceptable orbitals are available for the case at hand, and it is important to realize that in consequence there is no element either of arbitrariness or of ambiguity in the conclusions we shall reach.

We next note that the six electrons in the π orbitals of the product ethylene molecules must be assigned to the generalized molecular orbitals *(272)*–*(274)*.

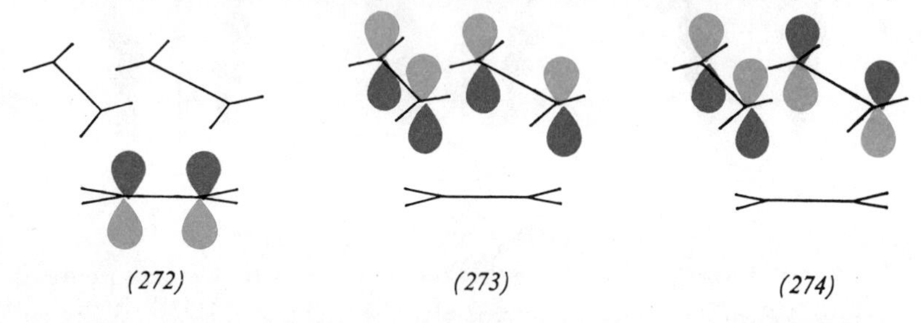

(272) *(273)* *(274)*

One further, and very important, factor must now be recognized. The geometrical relationships among starting material and products, as revealed by inspection of *(275)*, require that in the course of reaction, rotations about the axes 1,2 and 3,4

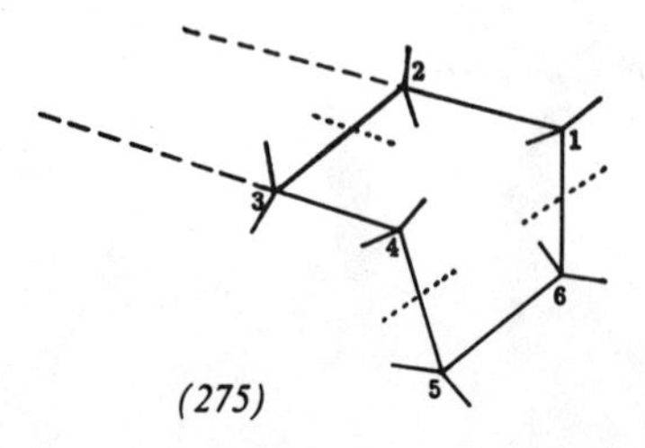

(275)

must take place, in such wise as to bring the substituent groups at C-1 and C-2 on the one hand, and those at C-3 and C-4 on the other, into the common planes in which they lie in the respective product ethylene molecules. *A priori*, on purely geometrical grounds, these displacements might be *conrotatory (276)* or *disrotatory (277)*, and there are, of course, in each case two possibilities [arrows in *(276)* and *(277)*].

(276) *(277)*

Now we are properly prepared to annihilate our cyclohexane molecule. We shall see that there are *two*, and only two, symmetry-allowed ways in which this decomposition may occur— given the transition state geometry we have chosen.

1. Let us first permit the two electrons in orbital *(265)* to enter orbital *(272)*, as of course they may do with conservation of orbital symmetry. It is then at once clear that the two electrons in orbital *(267)* must occupy orbital *(278)* of the product array — *and that of the rotational displacements adumbrated above, the inward disrotatory one (278)* must obtain. It remains only to place the remaining σ pair — those in the

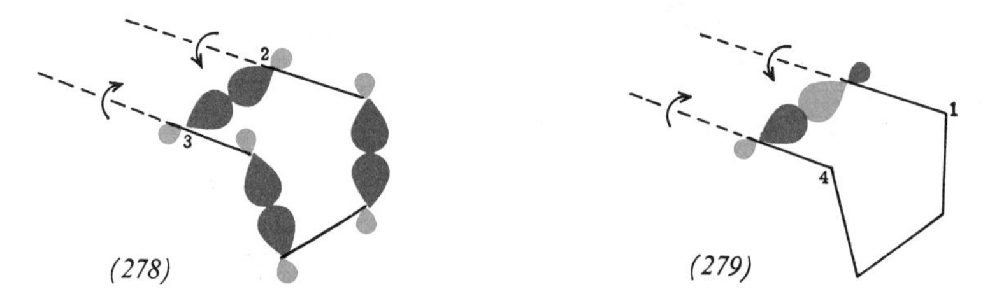

(278) *(279)*

orbital *(266)* at the outset of reaction — into orbital *(274)* of the product assemblage; a special point here is that mixing of the occupied orbital *(266)* with the unoccupied antibonding backbone σ^* orbital *(279)* is consistent with the same inward disrotatory motion about axes 1,2 and 3,4. The symmetry-allowed process which we have now depicted in full detail may be designated a —[$_{\pi}2_s+_{\pi}2_s+_{\pi}2_s$] cycloreversion reaction. Naturally, a precisely analogous analysis applies to the reverse process — the symmetry-allowed [$_{\pi}2_s+_{\pi}2_s+_{\pi}2_s$] cycloaddition reaction.

2. Alternatively, let us first permit the two electrons in orbital *(267)* to take up new positions in orbital *(272)* as cycloreversion proceeds. Now those in orbital *(265)* have no choice but to enter orbital *(273)* of the product array. And, in sharp contrast to the case discussed above, the cleavage of the backbone bond (2,3) must be accompanied by *outward* disrotatory displacements *(280)*. As before, the re-

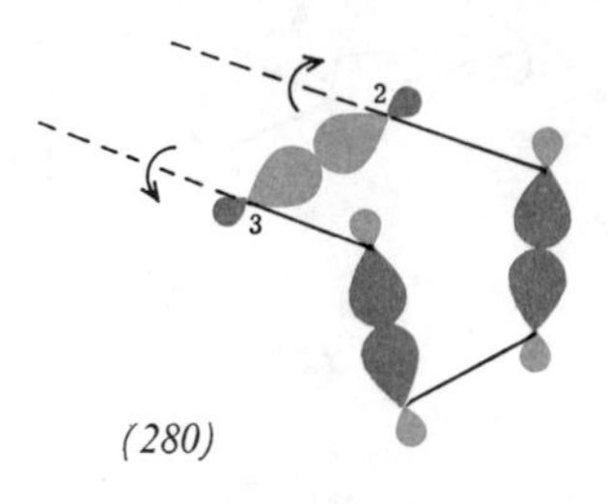

(280)

maining pair, occupying orbital *(266)*, takes up new positions in orbital *(274)*, and, as before, mixing of *(266)* with the unoccupied antibonding backbone σ^* orbital *(281)* is consistent with the outward disrotatory motion required by the symmetry relationships between orbitals *(280)* and *(274)*. The process here delineated may be designated a symmetry-allowed $-[_{\pi}2_s+_{\pi}2_a+_{\pi}2_a]$ cycloreversion reaction; it is the reverse of the allowed $[_{\pi}2_s+_{\pi}2_a+_{\pi}2_a]$ cycloaddition reaction.

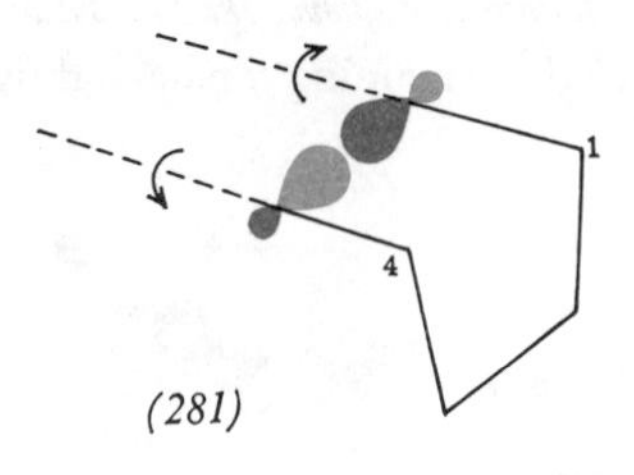

(281)

Thus, we have found that there are two possible symmetry-allowed [2+2+2] cyclic processes which proceed through a transition state possessing a single plane of symmetry. Structural, steric, and entropy factors may favor one or the other of these paths in various special circumstances. Is one or the other favored by inherent stereoelectronic factors not so far delineated? Indeed, when one considers in greater detail the mixing of the antisymmetric bonding orbital *(266)* with the antisymmetric antibonding orbital *(279)* [or *(281)*], it is found that the *s, s, s* process is favored over the alternative *s, a, a* path. Thus, the dihedral angle between the relevant orbitals at the outset of reaction is represented in plan in *(282)* and *(283)*.

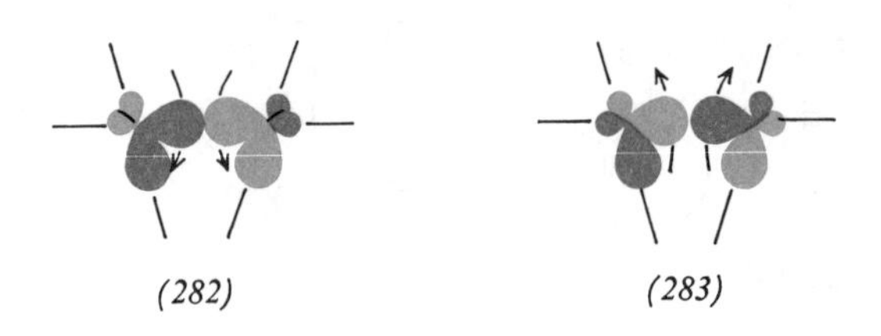

(282) (283)

The geometry of the alternatives indicates clearly that favorable, energy-lowering overlap will develop more readily from the motions depicted in *(282)*, and that the ascent to the transition state will be considerably steeper for the *s*, *a*, *a* process of *(283)*[128a].

Several facets of the [2+2+2] reactions deserve further, if now briefer, consideration. The reader may wish to sharpen his apprehension of symmetry considerations through verifying for himself the propositions which follow:

a) There is a continuum of topologically equivalent symmetry-allowed thermal $[_{\pi}2_s+_{\pi}2_s+_{\pi}2_s]$ processes, whose transition states vary from that depicted above to one whose geometry resembles that of a chair cyclohexane ring.

b) There are two further symmetry-allowed thermal $[_{\pi}2_s+_{\pi}2_a+_{\pi}2_a]$ processes, topologically different from the one described above, which proceed through enantiomeric transition states possessing a two-fold axis of symmetry.

c) There are *no* symmetry-allowed thermal $[_{\pi}2_s+_{\pi}2_s+_{\pi}2_a]$ processes.

d) There are *no* symmetry-allowed thermal $[_{\pi}2_a+_{\pi}2_a+_{\pi}2_a]$ processes.

e) For reactions involving occupied antibonding levels — for example, reactions of molecules in photochemically excited states — there are no *s*, *s*, *s* or *s*, *a*, *a* symmetry-allowed [2+2+2] processes, but there are two allowed enantiomeric *s*, *s*, *a*, as well as two permitted, and also enantiomeric, *a*, *a*, *a* paths.

f) Every all-*antara* [2+2+2+. . .] combination is symmetry-forbidden in ground states, and symmetry-allowed for excited states in which the antibonding orbital of one participating ethylene unit is occupied.

Lest any might cavil at the discussion of processes which on casual consideration could appear to be so improbable of realization as only to merit summary dismis-

[128a] Since the above was written, this prediction has received elegant experimental confirmation: *J. A. Berson* and *S. S. Olin*, J. Amer. chem. Soc. *91*, 777 (1969).

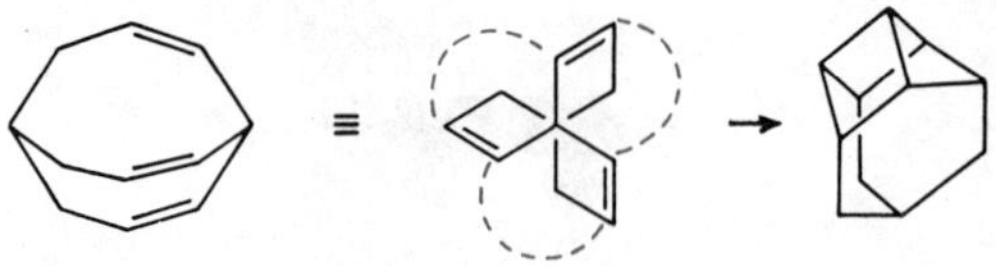

(284) (285)

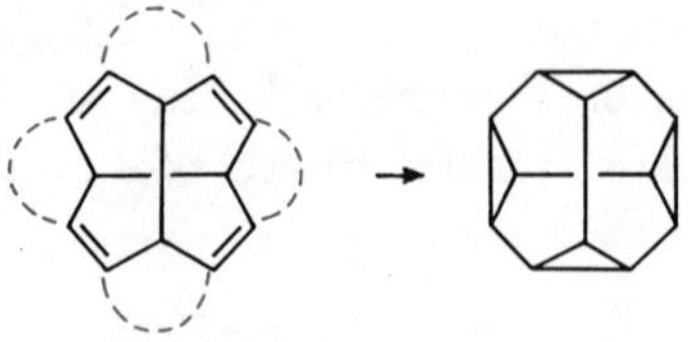

(286) (287)

sal, we may allude to the essentially strain-free, as yet unknown, molecules *(284)* and *(286)*, whose double bonds are so constrained by the frameworks in which they are embedded as to favor photochemical conversion to *(285)* and *(287)*, by the all-*antara* paths mentioned above.

It remains to exemplify some of the above processes through presentation of selected known cases. Most of those so far discovered fall in the symmetry-allowed thermal *s, s, s* class. Perhaps best known is the addition of olefins to bicycloheptadiene[129] [*(288)*→*(289)*].

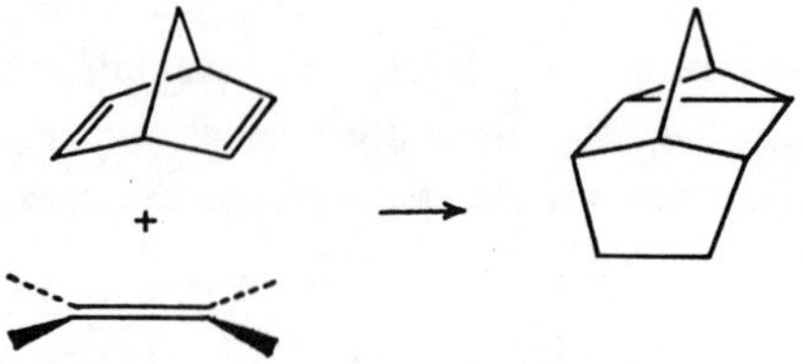

(288) (289)

Further examples in this group are the reactions *(290)*→*(291)*, *(292)*→*(293)*, and *(294)*→*(295)*.

[129] *Inter alia: A. T. Blomquist* and *Y. C. Meinwald*, J. Amer. chem. Soc. *81*, 667 (1959); *R. C. Cookson, S. S. H. Gilani*, and *I. D. R. Stevens*, Tetrahedron Letters *1962*, 615; *H. K. Hall*, J. org. Chem. *25*, 42 (1960).

(290)[130] *(291)*

(292)[131] *(293)*

(294)[132] *(295)*

6.4. Prismane

To what does this fantastic molecule *(296)* owe its capacity to survive? In it there is stored at least ninety kilocalories of energy more than is contained in its isomer benzene[133]; a graceful process of atomic displacements may readily be envisaged

(296)

---

[130] *J. K. Williams* and *R. E. Benson*, J. Amer. chem. Soc. *84*, 1257 (1962).
[131] *H. A. Staab, F. Graf*, and *B. Junge*, Tetrahedron Letters *1966*, 743.
[132] *C. D. Smith*, J. Amer. chem. Soc. *88*, 4273 (1966). *Cf.* also *H. Prinzbach*, Pure appl. Chem. *16*, 24 (1968); *H. Prinzbach* and *J. Rivier*, Angew. Chem. *79*, 1102 (1967); Angew. Chem. internat. Edit. *6*, 1069 (1967).
[133] *J. F. M. Oth*, Angew. Chem. *80*, 633 (1968); Angew. Chem. internat. Edit. *7*, 646 (1968); Rec. Trav. Chim. Pays-Bas *87*, 1185 (1968).

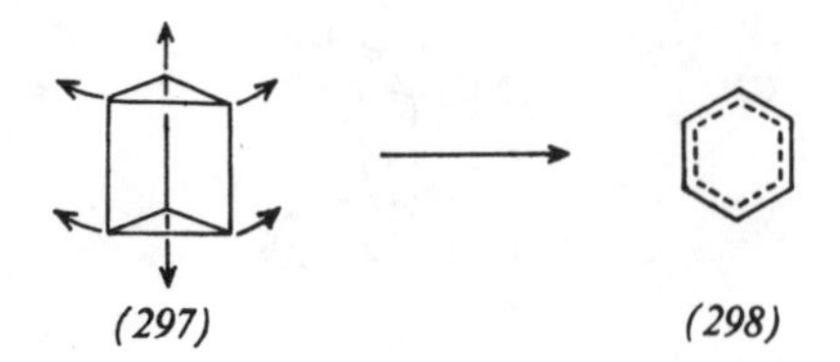

(297) (298)

for the conversion of the one molecule into the other [*cf.* *(297)*→*(298)*]. Yet, no less than a further thirty-three kilocalories must be added to the already exorbitantly strained hexamethylprismane molecule in order to bring about its conversion to hexamethylbenzene[133]. To the mind uninformed by an appreciation of the limitations placed upon bond-breaking and bond-making processes by orbital symmetry relationships, the excess energy of the prismane molecule must have the aspect of an angry tiger unable to break out of a paper cage.

When we examine this situation in detail from the vantage point of the principle of orbital symmetry conservation, we shall see that there is no paper cage, but rather

Antibonding Orbitals

S A A

A S A

(σ_1^*) (σ_2^*) (σ_3^*)

S A S

S S S

(σ_1) (σ_2) (σ_3)

Bonding Orbitals

(299)

one quite properly barred to contain its tiger. First, we must construct the generalized molecular orbitals *(299)* into which must be placed the six electrons of the three σ bonds which must be broken if prismane is to be converted to benzene.

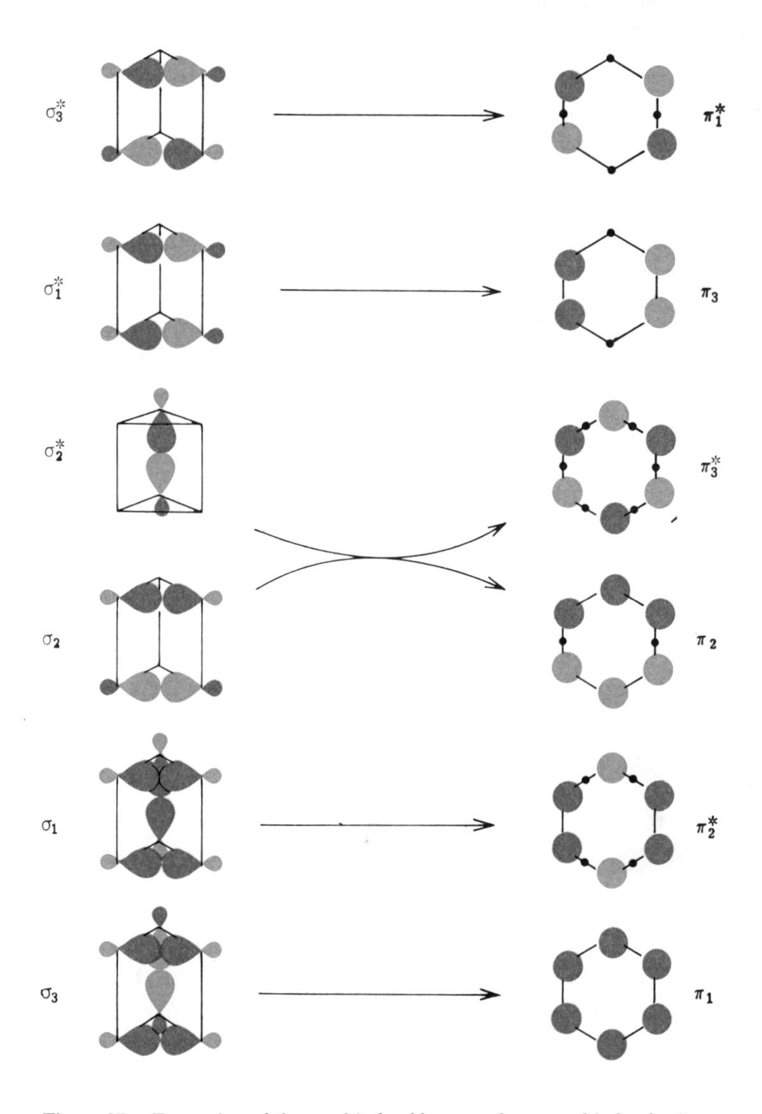

Figure 27. Formation of the π orbitals of benzene from σ orbitals of prismane.

Now, let us break the three σ bonds, *while conserving the symmetry imposed upon the orbitals of the prismane molecule by its geometry;* note that σ_2 and σ_2^*, of like symmetry, interact to produce a new pair, represented by the sum and the difference of the originals. The specified changes are shown in Figure 27.

What are the new orbitals thus generated? They are precisely the molecular orbitals of the benzene molecule. But while two of the bonding σ orbitals of prismane (σ_2 and σ_3) have transformed into two bonding orbitals of benzene (π_2 and π_1), *the third becomes an antibonding orbital of the product* ($\sigma_1 \rightarrow \pi_2^*$). Physically, this correlation introduces an antibonding, energy-raising component into the transition state for the transformation. It is that component which makes the transformation of prismane into benzene symmetry-forbidden, and permits the former to exist.

Further insight into the basic nature of symmetry-determined bonding relationships can be gained by amplifying our discussion of prismane in a number of ways. First, let us note that, in a formal sense, the transformation of prismane into ben-

(300) *(301)*

zene might be regarded as a $-[_{\pi}2_s + _{\pi}2_a + _{\pi}2_a]$ cycloreversion reaction [*(300)*$\rightarrow$ *(301)*]. We have seen in the preceding section that thermal reactions of that class are symmetry-allowed. What then excludes the prismane$\rightarrow$benzene transformation from the symmetry-allowed class to which it might formally be assigned on casual inspection? When we scrutinize the offending bonding-antibonding

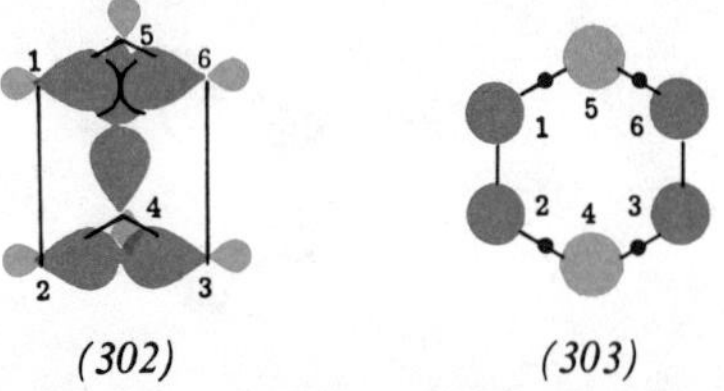

(302) *(303)*

correlation of the orbitals σ_1 *(302)* and π_2^* *(303)*, we see at once that it is the development of nodal, antibonding, energy-raising relationships at the 1,5, 5,6, 2,4, and 3,4 bonds, necessitated by the presence of the invariant framework σ bonds in the prismane molecule, which differentiates the prismane$\rightarrow$benzene transformation from the general [2+2+2] cycloreversion reaction. When, as in the latter, those σ bonds are absent, these antibonding, energy-raising components are also absent.

The point may be illustrated further by considering certain molecules closely related to prismane. Consider first quadricyclene *(304)* and the tricyclohexane *(305)*.

(304) *(305)*

These molecules might at first sight be expected to be susceptible of symmetry-allowed transformation to cycloheptatriene *(306)* and hexatriene *(307)*, but analysis

(306) *(307)*

along the lines set down above shows that, as in the case of prismane, the presence of the framework σ bonds places the changes in the symmetry-forbidden category. It is worthy of note that, in fact, quadricyclene is not converted to cycloheptatriene, even at very high temperatures[134]. In that light, the oxygen analogue *(308)* is of special interest. In this case, the product of a $-[_{\pi}2_s+_{\pi}2_a+_{\pi}2_a]$ cycloreversion is oxepine *(309)*, and the orbital correlation analogous to that which leads in the other

(308) *(309)*

cases to an antibonding level, gives now the highest occupied ground-state level, which is probably weakly bonding. It is gratifying to note that derivatives of the oxa analogue are smoothly transformed to oxepines when heated[135, 136].

---

[134] *H. Prinzbach* and *J. Rivier*, Tetrahedron Letter *1967*, 3713.
[135] *H. Prinzbach*, *M. Arguëlles*, and *E. Druckrey*, Angew. Chem. *78*, 1057 (1966); Angew. Chem. internat. Edit. *5*, 1039 (1966).
[136] *H. Prinzbach*, *P. Vogel*, and *W. Auge*, Chimia *21*, 469 (1967).

Another relative of the prismane molecule is the bicyclo[2.2.0]hexane *(310)*. In this case, lacking as it does σ bonds joining C-1 to C-5, and C-2 to C-4, no offensive nodes develop as the transition state for $-[_{\pi}2_s+_{\pi}2_a+_{\pi}2_a]$ cycloreversion is approached; consequently, the thermal transformation of *(310)* to ethylene and butadiene *(311)* is symmetry-allowed. In fact, it has not as yet been observed. When

(310) *(311)*

the unsubstituted bicyclohexane itself is heated, it is converted to 1,5-hexadiene *(312)*[137] — either by a two-step mechanism involving an intermediate *(313)*, or by a symmetry-allowed $[_{\sigma}2_s+_{\sigma}2_a]$ path.

(312) *(313)*

6.5. [2+2+2+2] Cycloadditions

The $[_{\pi}2+_{\pi}2+_{\pi}2+_{\pi}2]$ cycloaddition reaction is symmetry-allowed:

a) For ground state reactants, when an *odd* number of components is involved in a suprafacial manner.

b) For reactions in which one component is in an excited state, when an *even* number of components is involved in a suprafacial manner.

Obviously, entropy factors render polymolecular reactions of this type very highly improbable. Just as obviously, this difficulty can be surmounted by the manoeuvre of building all the components into a single molecule, geometrically prepared for their eventual dispersal. Such examples have already been mentioned above [*(284)* and *(286)*].

Symmetry-allowed many-component fragmentations are also opposed by a special factor; for each σ bond which is transformed into a π bond an endothermicity of some twenty kilocalories must be expected.

---

[137] *C. Steel, R. Zand, P. Hurwitz,* and *S. G. Cohen,* J. Amer. chem. Soc. *86,* 679 (1964).

There are instances of [$_\pi 2+_\pi 2+_\pi 2+_\pi 2$] cycloadditions masquerading as other processes. Consider for example the double cycloaddition of an acetylene to two ethylenes *(314)*. Examination of the orbitals reveals that the two acetylene π bonds are

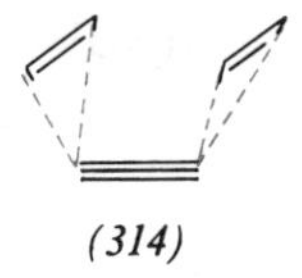

(314)

acting independently. At least two such reactions [*(315)*→*(316)* and *(317)*→*(318)*] have been observed[138, 139].

$H_3C—≡—CH_3$ hν CH_3 CH_3

(315) *(316)*

HOOC—≡—COOH hν HOOC COOH

(317) *(318)*

An amusing fragmentation would be the reaction *(319)*; the analogous decomposition of pentalene to diacetylene and two acetylenes has been suggested in the literature as a theoretical possibility[140]; we now recognize this to be a symmetry-forbidden thermal process.

(319)

[138] *R. Askani*, Chem. Ber. *98*, 3618 (1965).
[139] *M. Takahashi, Y. Kitahara, I. Murata, T. Nitta,* and *M. C. Woods*, Tetrahedron Letters *1968*, 3387.
[140] *H. C. Longuet-Higgins* in: Theoretical Organic Chemistry. The Kekulé Symposium. Butterworths, London 1959, p. 17.

7. Theory of Sigmatropic Reactions

We defined as a sigmatropic change of order [i,j] the migration of a σ bond, flanked by one or more π electron systems, to a new position whose termini are i-1 and j-1 atoms removed from the original bonded loci, in an uncatalyzed intramolecular process. Thus, the well-known Claisen and Cope rearrangements are sigmatropic changes of order [3,3].

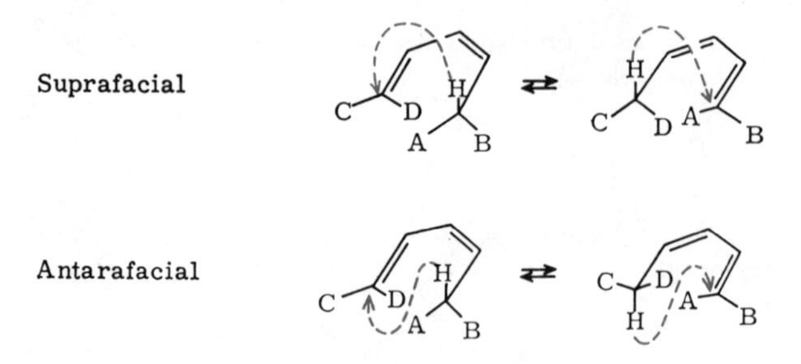

Figure 28. Suprafacial and antarafacial [1,5] shifts of a hydrogen atom.

A priori, there are two topologically distinct ways of effecting a sigmatropic migration. These are illustrated in Figure 28 for the [1,5] shifts of a hydrogen atom. In the first, *suprafacial* process, the transferred hydrogen atom is associated at all times with the same face of the π system. In the second, *antarafacial* process, the migrating atom is passed from the top face of one carbon terminus to the bottom face of the other.

For the analysis of these reactions correlation diagrams are not relevant since it is only the transition state and not the reactants or products which may possess molecular symmetry elements. We shall present several equivalent methods for analyzing these reactions.

1. The use of the principle of *conservation of orbital symmetry* is here illustrated for the case of a suprafacial [1,3] hydrogen shift. The relevant correlations are shown in Figure 29. Clearly, two electrons can enter a bonding orbital, either σ or π, of the product, but the other two must be placed either in a σ^* or a π^* orbital, if orbital symmetry is to be conserved. *The reaction is symmetry-forbidden.*

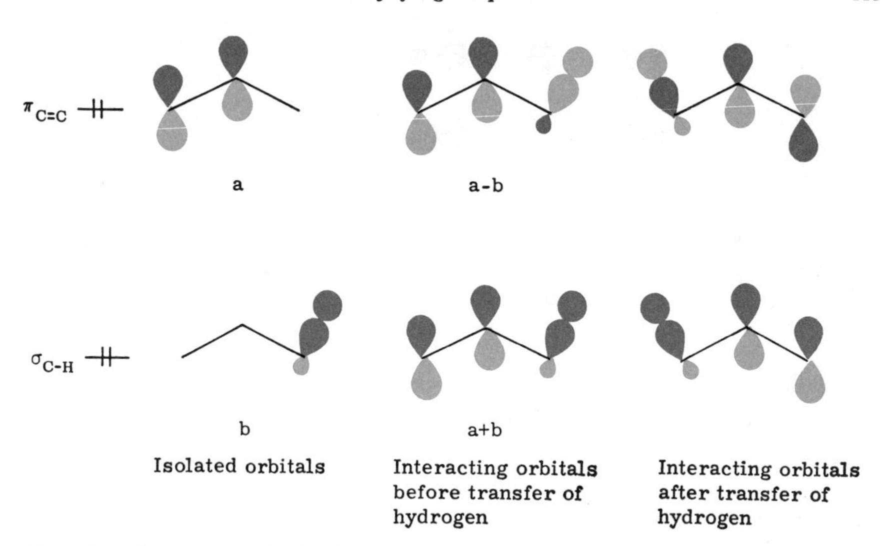

Figure 29. Conservation of orbital symmetry in a suprafacial [1,3] hydrogen shift.

2. In the [1,j] sigmatropic migration of hydrogen within an all-*cis*-polyene framework [*(320)*→*(321)*], one may envisage the transition state as made up by the combination of the orbital of a hydrogen atom with those of a radical containing

$$R_2\overset{1}{C}=\overset{2}{C}H-(CH=CH)_k-\overset{j}{C}HR_2 \rightarrow R_2\overset{1}{C}H-(CH=CH)_k-CH=CR_2$$

(320) *(321)*

$2k+3$ π electrons. The highest occupied orbital of the framework system is the nonbonding allylic orbital which possesses the symmetry shown in *(322)*. Consider a hydrogen atom bonding to the termini of such systems. In order to avoid a high energy orbital in the transition state, it is required that positive overlap

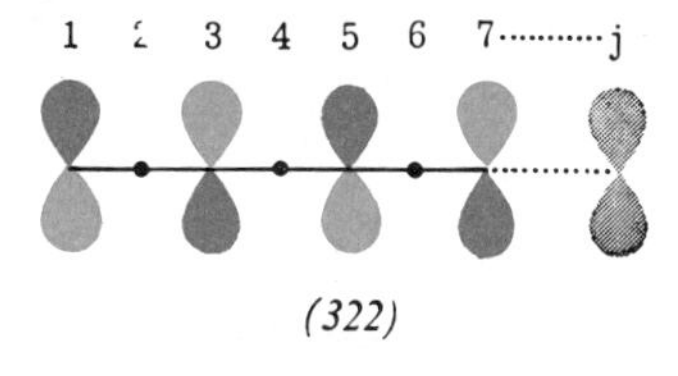

(322)

between the framework orbital and the migrating hydrogen orbital be maintained. The result is symmetry-allowed suprafacial migration for odd k, while the antarafaci l path must be followed for even k.

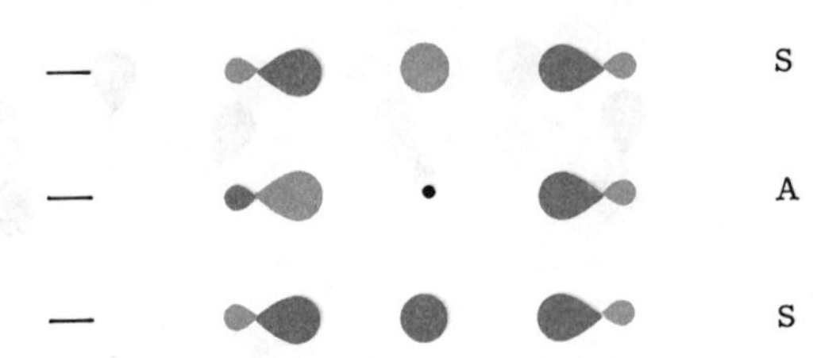

Figure 30. Orbital relationship in a three-center bond. Note that the symmetry labels are invariant whether a mirror plane or a two-fold axis of rotation is used as the basis for classification.

3. Let us now examine the transition state from a different, but equivalent, viewpoint. First, we write down an explicit three-center bond involving the termini of the carbon chain and the migrating hydrogen atom. The three-center bond has the typical level pattern indicated in Figure 30. There remains a $2k+1$ π orbital system in the carbon chain. The molecular orbitals of this remnant are next classified with respect to the symmetry element (two-fold axis or mirror plane) present in the transition state. There now arise two cases, differing in the symmetry of the non-

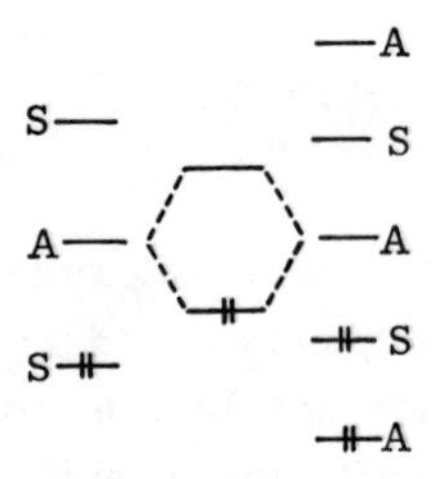

Figure 31. Interaction of nonbonding antisymmetric levels of three-center bond and remnant, for the antarafacial [1,7] shift of a hydrogen atom ($k = 2$).

bonding orbital of the remnant. In the first case (Figure 31) this orbital is antisymmetric and interacts strongly with the central orbital of the three-center bond. In the second case (Figure 32) the remnant orbital is symmetric and cannot interact in a similar way. The presence of the interaction typically stabilizes one level, and destabilizes the other. Now, since there are in all $2k+4$ electrons to be accommodated — $2k$ electrons in bonding polyenyl levels, and 2 in the lowest orbital of the three-

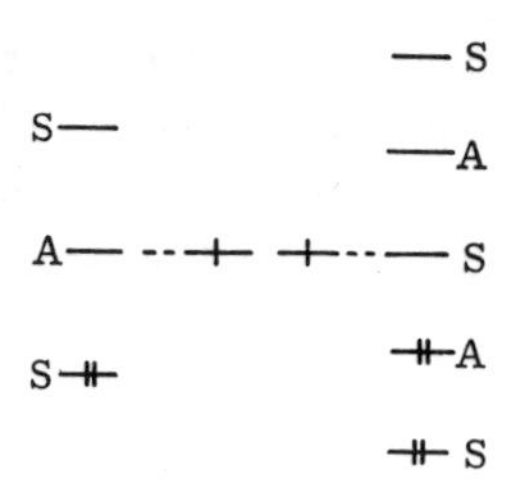

Figure 32. Noninteraction of nonbonding levels of three-center bond and remnant, for the suprafacial [1,7] shift of a hydrogen atom ($k = 2$).

center bond — there remain 2 electrons to be placed in the lower component of this interacting pair. It follows that for the ground-state reaction, the first case is symmetry-allowed. Thus, in general, in order that the nonbonding orbital of the $(2k+1)$-orbital remnant π system shall be antisymmetric under the appropriate symmetry operation, these reactions must proceed through a transition state having a mirror plane when k is odd, or one having a two-fold axis when k is even.

For excited-state reactions, as in the cases of electrocyclic reactions and cycloadditions, the selection rules are precisely reversed, as compared with those for ground states.

For sigmatropic changes of order [i,j] in which both i and j are greater than 1, the migrating group of the above analyses consists of more than one atom, and appropriate topological distinctions must be made in relation to *both* π systems through which the σ bond is moving. The selection rules are summarized in Figure 33.

$i + j$	Ground State	Excited State
$4q$	antara-supra	supra-supra
	supra-antara	antara-antara
$4q + 2$	supra-supra	antara-supra
	antara-antara	supra-antara

Figure 33. Selection rules for sigmatropic reactions of order [i,j] with i and $j > 1$.

In the above treatment of [1,j] sigmatropic changes, we have considered only those cases in which a σ orbital of the migrating group interacts with a π system in the transition state, and migration occurs *with retention of configuration at the shifting site*. When the migrating group possesses an accessible π orbital and is not so substituted as to create an impossible steric situation in the transition state, alternative processes using that π orbital must be considered. Clearly, such changes *must proceed with inversion at the migrating center* [*cf.* *(323)*], and the selection rules are precisely reversed from those given in Figure 33.

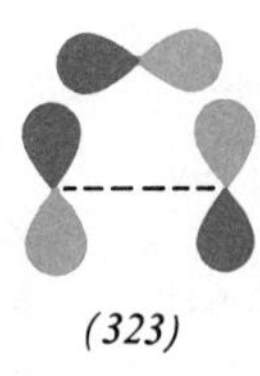

(323)

A number of special points pertinent to sigmatropic changes should be noted.

a) Antarafacial processes are obviously impossible for transformations which occur within small or medium-sized rings.

b) Distortion of the carbon framework, with concomitant impairment of coupling within the π system, may render a symmetry-allowed process difficult or impossible of realization; for example, this factor makes the antarafacial process difficult or perhaps impossible in the [1,3] case, but is not an impediment for the [1,7] reaction.

c) A cyclopropane ring may replace a π bond in the framework system for a sigmatropic change.

d) Orbital symmetry arguments are applicable to sigmatropic changes within ionic species. Thus, the suprafacial [1,2] shift within a carbonium ion is symmetry allowed and is very well known. The predicted [1,4] shift with inversion of the migrating group, within a but-2-en-1-yl cation has recently been observed. We may expect that the [1,6] shift within a hexa-2,4-dien-1-yl cation will take place through a readily accessible suprafacial transition state.

It remains to point out that the insights afforded by the discussions presented in the preceding sections allow us to view sigmatropic changes as cycloaddition reactions. Thus, the [1,3] sigmatropic process *(324)*→*(325)* is a $[_{\sigma}2+_{\pi}2]$ reaction, and

R

R

(324) *(325)*

the [3,3] sigmatropic rearrangement *(326)*→*(327)* is a $[_{\pi}2+_{\sigma}2+_{\pi}2]$ concerted cycloaddition.

(326) *(327)*

From this standpoint, let us now examine in detail the stereochemical aspects of the [1,3] sigmatropic change. We know that the symmetry-allowed processes must fall in either the $[_{\sigma}2_s+_{\pi}2_a]$ or the $[_{\sigma}2_a+_{\pi}2_s]$ class. All of the possibilities are shown in Figure 34. It is at once apparent that the first two alternatives depicted in

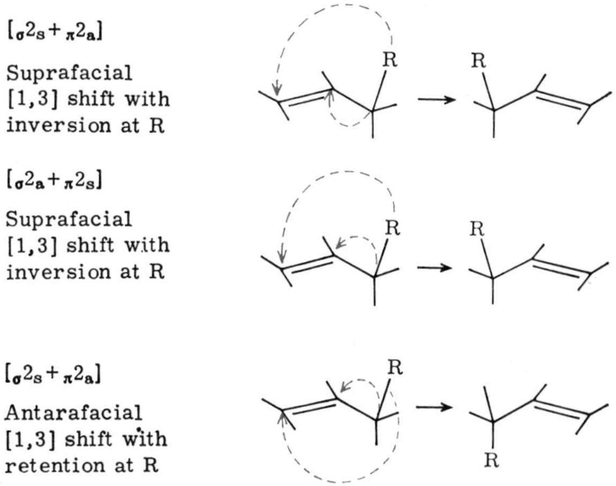

$[_{\sigma}2_s+_{\pi}2_a]$

Suprafacial
[1,3] shift with
inversion at R

R R

$[_{\sigma}2_a+_{\pi}2_s]$

Suprafacial
[1,3] shift with
inversion at R

R R

$[_{\sigma}2_s+_{\pi}2_a]$

Antarafacial
[1,3] shift with
retention at R

R R

Figure 34. Stereochemical possibilities in sigmatropic [1,3] shifts.

Figure 34 are simply different ways of regarding one and the same physical process. Consequently, the analysis yields, as it must, precisely the conclusions reached in our prior discussions of sigmatropic reactions.

7.1. Sigmatropic Reactions Exemplified

Few thermal uncatalyzed [1,3] shifts are known[141]. Those which formally might be considered in this category, such as the vinylcyclopropane→cyclopentene rearrangement, proceed with such high activation energies (*ca.* 50 kcal/mole) that the energy surface for concerted reaction cannot be far removed from that for alternative stepwise processes. None the less, *Berson* and *Nelson*[142] recently made the dramatic observation that *(328)* undergoes a concerted symmetry-allowed suprafacial [1,3]shift, ***with inversion at the migrating center*** [starred in *(328)*], to give *(329)*, at 307° C.

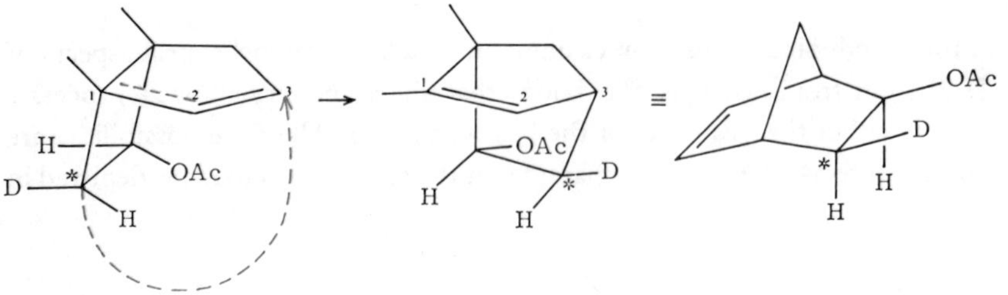

(328) *(329)*

The triene *(330)* owes its relative stability[143] to the requirement of a symmetry-forbidden [1,3] shift for an uncatalyzed rearrangement to toluene.

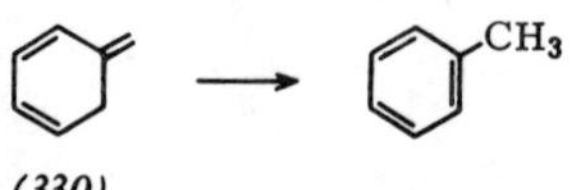

(330)

The rearrangement of the ester *(331)* of Feist's acid is an especially interesting [1,3] sigmatropic transformation. Studies on the optically active substance have demonstrated that the reaction occurs with high stereospecificity[144]. Of the two *a*

---

[141] For example, [1-^{14}C]-propylene does not rearrange to [3-^{14}C]-propylene: *B. Sublett* and *N. S. Bowman*, J. org. Chem. *26*, 2594 (1961).

[142] *J. A. Berson* and *G. L. Nelson*, J. Amer. chem. Soc. *89*, 5303 (1967); *J. A. Berson*, Accounts Chem. Res. *1*, 152 (1968).

[143] *W. J. Bailey* and *R. A. Baylouny*, J. org. Chem. *27*, 3476 (1962).

[144] *W. von E. Doering*, personal communication; *cf. E. F. Ullman*, J. Amer. chem. Soc. *82*, 505 (1960).

priori possible symmetry-allowed skeletal processes — [1,3] suprafacial migration with inversion at the migrating atom, and [1,3] antarafacial shift with retention — the latter is rendered inaccessible by the constraints imposed by the σ skeleton.

Since substituents are present, the allowed and accessible process may be realized in two electronically identical, but sterically different, ways. Consequently, we may predict that *(331)* must be transformed into *(332)* and/or *(333)*; the evidence is now conclusive[144] that the two isomers actually produced are the predicted ones.

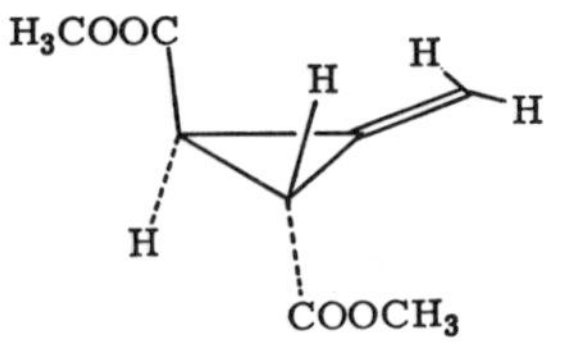

(331) 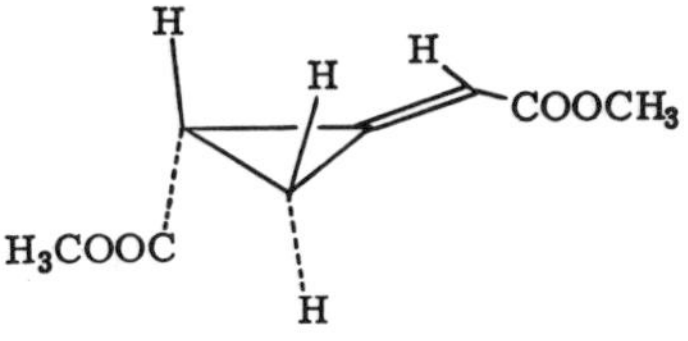 *(332)*

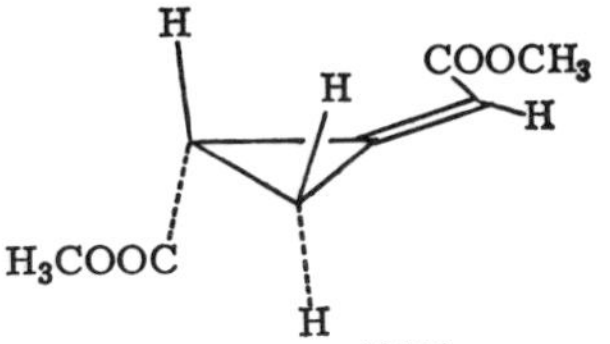

(333)

We have mentioned above that the activation energy for the vinylcyclopropane→cyclopentene conversion[145, 146] is *ca.* 50 kcal/mole[146]. The bond-breaking process in cyclopropane itself requires an activation energy of 63 kcal/mole[147]; allylic stabilization has been estimated at *ca.* 13 kcal/mole[148]. Consequently, if full use be made of the available allylic stabilization energy in a diradical intermediate, a two step, non-concerted path for the conversion of vinylcyclopropane to cyclo-

---

[145] *C. G. Overberger* and *A. E. Borchert*, J. Amer. chem. Soc. *82*, 1007 (1960).
[146] *M. C. Flowers* and *H. M. Frey*, J. chem. Soc. *1961*, 3547; *R. J. Ellis* and *H. M. Frey*, ibid. *1964*, 959, 4188; *C. J. Elliot* and *H. M. Frey*, ibid. *1965*, 345; *A*, *1966*, 553; *C. A. Wellington*, J. physic. Chem. *66*, 1671 (1962); *H. M. Frey* and *D. C. Marshall*, J. chem. Soc. *1962*, 3981; *G. R. Branton* and *H. M. Frey*, ibid. *1966*, 1342.
[147] *B.S. Rabinowitch*, *E. W. Schlag*, and *K. B. Wiberg*, J. chem. Physics *28*, 504 (1958); *B.S. Rabinowitch* and *E. W. Schlag*, J. Amer. chem. Soc. *86*, 5996 (1960).
[148] *K. W. Egger*, *D. M. Golden*, and *S. W. Benson*, J. Amer. chem. Soc. *86*, 5420 (1964).

pentene is not thermodynamically unreasonable. Some preliminary studies have yielded results consistent with such a mechanism[149]. Nevertheless, in view of the recently established occurrence of symmetry-allowed concerted processes, despite extraordinarily unfavorable geometric constraints, it is worth while to consider the stereochemical consequences of the possible concerted processes for the vinylcyclopropane→cyclopentene change. Consider the maximally labeled derivative *(334)*; the symmetry-allowed reactions are:

a) an antarafacial [1,3] shift of bond 1,2 to C-5, with retention at C-2, leading to *(335)*;

b) a suprafacial [1,3] shift of bond 1,2 to C-5, with inversion at C-2, leading to *(336)*.

The isomer *(337)* cannot be produced in a symmetry-allowed process.

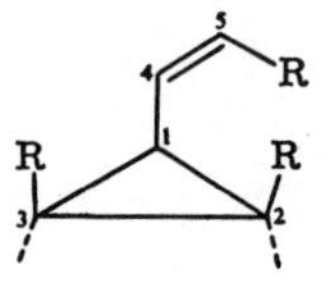

(334) 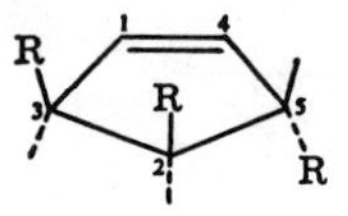 *(335)* 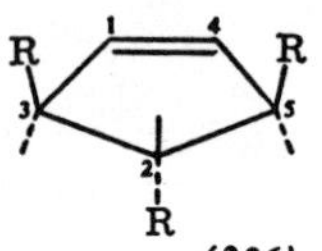 *(336)*

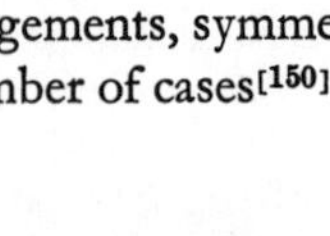

(337)

Photochemical [1,3] rearrangements, symmetry-allowed as suprafacial shifts, have been observed in a large number of cases[150].

---

[149] *M. R. Willcott* and *V. M. Cargle*, J. Amer. chem. Soc. *89*, 723 (1967); *W. R. Roth*, personal communication.

[150] *W. G. Dauben* and *W. T. Wipke*, Pure appl. Chem. *9*, 539 (1964); *J. J. Hurst* and *G. M. Whitham*, J. chem. Soc. *1960*, 2864; *R. C. Cookson, V. N. Gogte, J. Hudec*, and *N. A. Mirza*, Tetrahedron Letters *1965*, 3955; *W. F. Erman* and *H. C. Kretschmar*, J. Amer. chem. Soc. *89*, 3842 (1967); *R. F. C. Brown, R. C. Cookson*, and *J. Hudec*, Tetrahedron *24*, 3955 (1968); *R. C. Cookson*, Chem. in Britain *5*, 6 (1969); *E. Baggiolini, H. P. Hamlow, K. Schaffner* and *O. Jeger*, Chimia *23*, 181 (1969).

Specific thermal [1,5] shifts have been observed in a great number of cases[151]. The simplest case of a [1,5] shift in a 1,3-pentadiene was studied by *Roth* and *König*[152]. They compared the rearrangements of *(338)* and *(339)*, and observed a large kinetic isotope effect of 12.2 at 25° C, consistent with a highly symmetrical transition state in a concerted process. The activation energy for the rearrange-

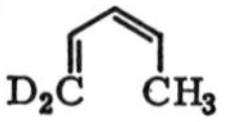

D_2C CH_3 H_2C CD_3

(338)  *(339)*

ment is in the neighborhood of 35 kcal/mole. Direct confirmation of the concerted, suprafacial character of the [1,5] sigmatropic shift of a hydrogen atom has been most elegantly provided by the same investigators[153], who demonstrated the completely stereospecific conversion of *(340)* into *(341)* and *(342)*.

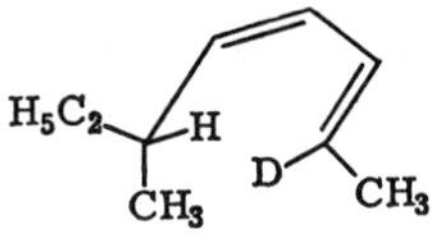

H_5C_2 H CH_3 D CH_3

(340)

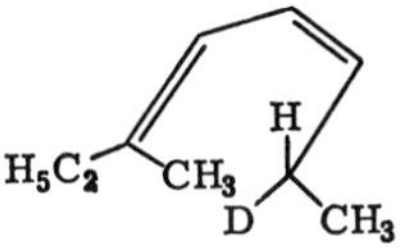

H H_5C_2 CH_3 D CH_3

(341)

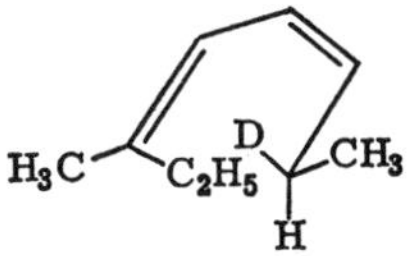

D H_3C C_2H_5 CH_3 H

(342)

Numerous sequential stereospecific [1,5] shifts have been observed in cyclopentadienes and cycloheptatrienes[154—156].

[151] References to the literature may be found in *D. S. Glass, R. S. Boikess*, and *S. Winstein*, Tetrahedron Letters *1966*, 999. An interesting recently discovered [1,5] shift of a methyl group is described by *R. Grigg, A. W. Johnson, K. Richardson*, and *K. W. Shelton*, Chem. Commun. *1967*, 1192; *cf.* also *V. Boekelheide* and *E. Sturm*, J. Amer. chem. Soc. *91*, 902 (1969).
[152] *W. R. Roth* and *J. König*, Liebigs Ann. Chem. *699*, 24 (1966). See also *H. Kloosterziel* and *A. P. ter Borg*, Rec. Trav. Chim. Pays-Bas *84*, 1305 (1965).
[153] *W. R. Roth* and *J. König*, personal communication.
[154] *V. A. Mironov, E. V. Sobolev*, and *A. N. Elizarova*, Tetrahedron *19*, 1939 (1963); *S. McLean* and *R. Haynes*, Tetrahedron Letters *1964*, 2385.
[155] *W. R. Roth*, Tetrahedron Letters *1964*, 1009.
[156] *A. P. ter Borg, H. Kloosterziel*, and *N. van Meurs*, Rec. Trav. Chim. Pays-Bas *82*, 717, 741, 1189 (1963); *E. Weth* and *A. S. Dreiding*, Proc. chem. Soc. *1964*, 59; *K. W. Egger*, J. Amer. chem. Soc. *89*, 3688 (1967).

Roth has beautifully shown the absolute preference for [1,5] over [1,3] shifts in cyclic systems. Thus, a [1,5] shift scrambles the deuterium label in *(343)* over all nonaromatic positions at elevated temperatures[155], despite the necessity of proceeding through the unstable isoindene *(344)*. In sharp contrast, the base-catalyzed reaction proceeds with a stereospecific [1,3] shift[157].

(343) ⇄ *(344)*

Still another test for the possible occurrence of a thermal [1,3] shift proved negative[158]. The 7,8 dideuterium labeled cycloocta-1,3,5-triene *(345)*, through a series of reversible [1,3] sigmatropic shifts, involving the isomeric cycloocta-1,3,6-triene, would distribute the label statistically over all ring positions. On the other hand, a sequence of [1,5] shifts would place the label in positions 3, 4, 7, and 8 only; the latter result is experimentally observed.

[1,3] ⇄ *(345)* [1,5] ⇄

Photochemical [1,5] shifts of as yet undetermined stereochemistry are known in open, but *not* in cyclic hydrocarbons[159] and in heterocycles[160].

Berson and *Willcott*[161] have observed a remarkable series of isomerizations of methyl-substituted cycloheptatrienes. Their results, summarized in Figure 35, are

[157] *G. Bergsson* and *A. Weidler*, Acta chem. scand. *17*, 1798 (1963); *A. Weidler*, ibid. *17*, 2724 (1963); *A. Weidler* and *G. Bergsson*, ibid. *18*, 1484, 1487 (1964); *J. Almy*, *R. T. Uyeda*, and *D. J. Cram*, J. Amer. chem. Soc. *89*, 6768 (1967), and references therein.
[158] *W. R. Roth*, Liebigs Ann. Chem. *671*, 25 (1964).
[159] For example: *R. Srinivasan*, J. Amer. chem. Soc. *84*, 3982 (1962); *K. J. Crowley*, Proc. chem. Soc. *1964*, 17; *H. Prinzbach* and *E. Druckrey*, Tetrahedron Letters *1964*, 2959.
[160] *E. F. Zwicker*, *L. I. Grossweiner*, and *N. C. Yang*, J. Amer. chem. Soc. *85*, 2671 (1963); *G. Wettermark*, Photochem. Photobiol. *4*, 621 (1965); *K. R. Huffman*, *M. Loy*, and *E. F. Ullman*, J. Amer. chem. Soc. *87*, 5417 (1965); *G. Büchi* and *N. C. Yang*, ibid. *79*, 2318 (1957).

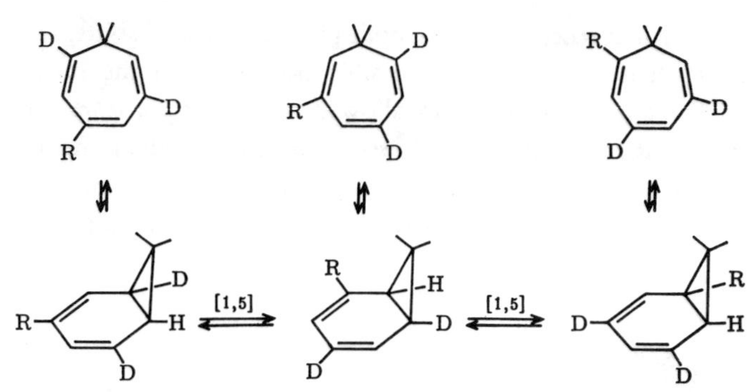

Figure 35. Isomerizations of methyl-substituted cycloheptatrienes through sequences of [1,5] sigmatropic rearrangements and electrocyclic reactions.

consistent with a sequence of [1,5] sigmatropic rearrangements and electrocyclic reactions. Our prediction is that the suprafacial [1,5] shifts must proceed with retention at the shifting center. This condition implies that in each shift substituents on the cyclopropane ring pivot 180°. The beautiful experimental test proposed but as yet not completed by *Berson* and *Willcott*[161] necessitates the pyrolysis of an optically active compound *(346)*. Symmetry-forbidden [1,5] shifts with inversion —

A B

R

(346)

the course which would be followed were the principle of least motion determinative — would result in racemization, while the predicted shifts with retention would lead to no loss of optical activity (Figure 36).

Figure 36. Schematic top view of the possible placements of substituents on a cyclopropane ring migrating around a cyclohexadiene ring by a series of [1,5] sigmatropic shifts, with retention (left) or inversion (right). Note that for clarity the cyclopropane rings are assumed to be perpendicular to the plane of the page, and are not seen.

---

[161] *J. A. Berson* and *M. R. Willcott, III*, J. Amer. chem. Soc. *87*, 2751, 2752 (1965); *88*, 2494 (1966).

The Cope and Claisen rearrangements, both [3,3] sigmatropic shifts, are such common reactions that we could not possibly do justice to the many examples studied[162]. These rearrangements are allowed both in a *supra-supra* and an *antara-antara* manner. At first sight it would seem obvious that the transition state for the latter is sterically unattainable. And yet an *antara-antara* Cope rearrangement has been observed; *Mukai*[163] reported the stereospecific rearrangement of *(347)*

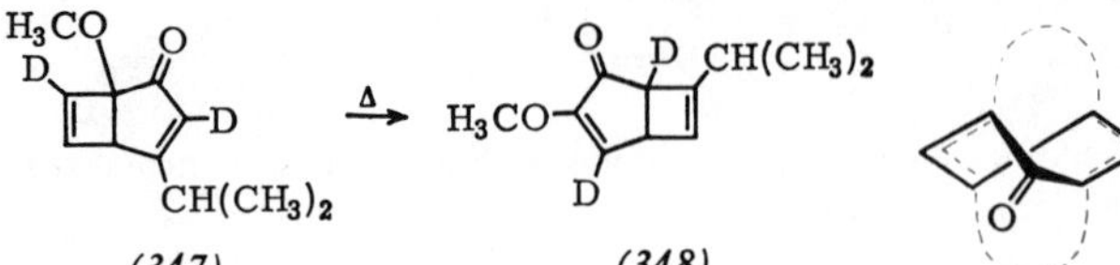

(347) *(348)* *(349)*

to *(348)*. The case provides a striking instance of the manner in which framework restraints may operate to make otherwise difficultly accessible symmetry-allowed reactions possible [*cf.* the transition state *(349)*].

The antarafacial sigmatropic transformation of order [1,7] has been known for some time, in the precalciferol→calciferol interconversion[164].

In cycloheptatriene an antarafacial [1,7] shift is impossible. Consequently, [1,7] shifts within this system must be photochemically induced, or occur with inversion at the shifting site if thermal. A number of sequences of such light-induced [1,7] rearrangements have been observed[165].

Murray and *Kaplan*[166] have observed the consistent sequence *(350)*—*(353)* of ther-

---

[162] *Cf. W. von E. Doering* and *W. R. Roth*, Tetrahedron *18*, 67 (1962); Angew. Chem. *75*, 27 (1963); Angew. Chem. internat. Edit. *2*, 115 (1963); *S. J. Rhoads* in *P. de Mayo:* Molecular Rearrangements. Interscience, New York 1963, Vol. 1, p. 655; *A. Jefferson* and *F. Scheinmann*, Quart. Rev. *22*, 391 (1968).
[163] *T. Miyashi, M. Nitta*, and *T. Mukai*, Tetrahedron Letters *1967*, 3433.
[164] *J. L. M. A. Schlatmann, J. Pot*, and *E. Havinga*, Rec. Trav. Chim. Pays-Bas *83*, 1173 (1964); *M. Akhtar* and *C. J. Gibbons*, Tetrahedron Letters *1965*, 509.
[165] *A. P. ter Borg* and *H. Kloosterziel*, Rec. Trav. Chim. Pays-Bas *84*, 241 (1965); *W. von E. Doering* and *P. P. Gaspar*, J. Amer. chem. Soc. *85*, 3043 (1963); *W. R. Roth*, Angew. Chem. *75*, 921 (1963); Angew. Chem. internat. Edit. *2*, 688 (1963); *L. B. Jones* and *V. K. Jones*, J. Amer. chem. Soc. *90*, 1540 (1968).
[166] *R. W. Murray* and *M. L. Kaplan*, J. Amer. chem. Soc. *88*, 3527 (1966).

mal [1,5] and photochemical [1,7] sigmatropic hydrogen shifts among the isomers of 1,4-di(cycloheptatrienyl)benzene.

(350)

[1,7] hν [1,5] Δ

(351) Δ [1,5] (352)

[1,5] Δ hν [1,7]

(353)

Further fascinating examples of [1,7] shifts have been described by *Schmid*[166a].

A symmetry-allowed [3,5] sigmatropic rearrangement must be an excited state reaction if constrained to be suprafacial on both components. So far, there appears to have been observed one case only[167].

The first [5,5] sigmatropic shift has now been realized by *Fráter* and *Schmid*[168], in the facile stereospecific rearrangement of *(354)* to *(355)*.

(354) Δ (355)

[166a] *R. Hug, H.-J. Hansen*, and *H. Schmid*, Chimia *23*, 108 (1969).
[167] *K. Schmid* and *H. Schmid*, Helv. chim. Acta *36*, 687 (1953).
[168] *Gy. Fráter* and *H. Schmid*, Helv. chim. Acta *51*, 190 (1968).

[1,2] Shifts in a carbonium ion are very common[169]. Stereospecific hydrogen or methyl shifts in benzenonium cations[170] may be regarded as [1,2] or [1,6] shifts.

Sigmatropic reactions of order [1,4], *e.g.* migrations in a but-2-en-1-yl cation, have not been observed until recently, when *Swatton* and *Hart* reported the very rapid degenerate rearrangement *(356)*[171].

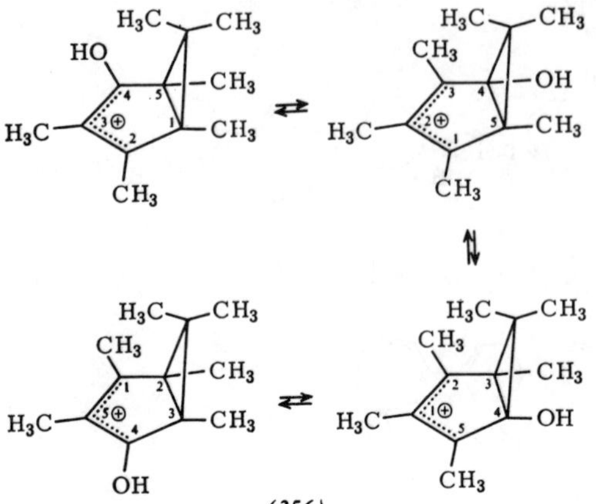

(356)

A [1,4] sigmatropic shift constrained to proceed suprafacially must take place with inversion at the migrating group. Note how this prediction contrasts with that for the [1,5] shift in the sterically very similar *(357)*.

(357)

[169] *Cf.* the reviews by *Y. Pocker*, and by *J. A. Berson* in *P. de Mayo:* Molecular Rearrangements. Interscience, New York 1963, Vol. 1.
[170] *D. A. McCaulay* and *A. P. Lien*, J. Amer. chem. Soc. *79*, 5953 (1957); *H. Steinberg* and *F. L. J. Sixma*, Rec.Trav. Chim. Pays-Bas *81*, 185 (1962); *C. MacLean* and *E. L. Mackor*, J. chem. Physics *34*, 2208 (1961); *V. A. Koptyug, V. G. Shubin*, and *A. I. Rezvukhin*, Izv. Akad. Nauk SSSR *1965*, 201.
[171] *D. W. Swatton* and *H. Hart*, J. Amer. chem. Soc. *89*, 5075 (1967).

The predicted inversion in a [1,4] sigmatropic shift has recently been verified by several groups. *Hill*[172], and *Zimmerman*[173] have demonstrated that such a process occurs in the Favorskii reaction of a bicyclo[3.1.0]hexanone [*(358)*→*(359)*→*(360)*].

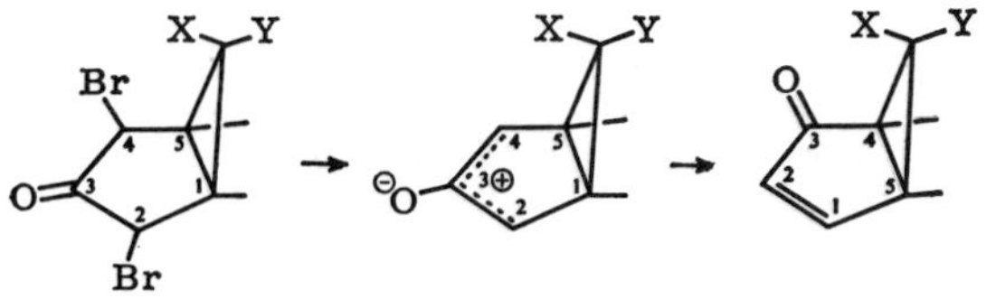

(358) *(359)* *(360)*

Further, *Hart*[174] has demonstrated that the rearrangement *(356)* previously discovered by him[171] indeed proceeds with inversion. The clearest demonstration of the nature of this rearrangement has been obtained by *Winstein* and *Childs*[175]. They generated various methylated bicyclo[3.1.0]hexenyl cations *(362)* in highly acidic

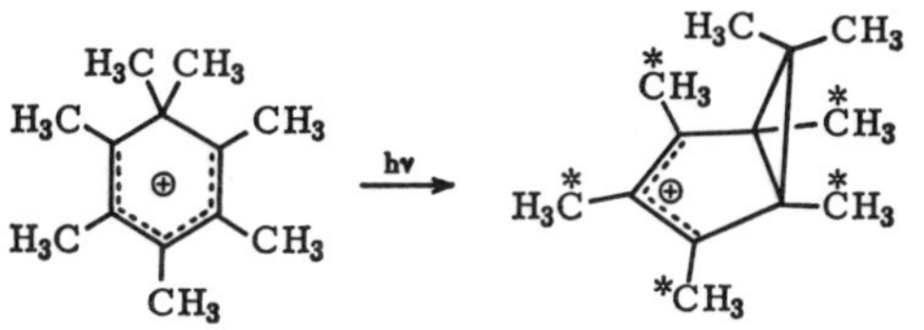

(361) *(362)*

media under conditions such that the nuclear magnetic resonance spectrum of the cation could be observed directly at various temperatures. The cations were generated by the symmetry-allowed photochemical disrotatory cyclization of the corresponding cyclohexadienyl cations *(361)* and were relatively stable, in consequence of the symmetry-imposed barrier to thermal reversion. As the temperature at which the measurements were made was raised, in the case of the heptamethyl compound *(362)* an averaging process took place; five methyl groups (marked by asterisks) became equivalent, while the other two methyl groups, which at low temperature gave distinct resonances, remained unaffected.

---

[172] *T. M. Brennan* and *R. K. Hill*, J. Amer. chem. Soc. *90*, 5614 (1968).
[173] *H. E. Zimmerman* and *D. S. Crumrine*, J. Amer. chem. Soc. *90*, 5612 (1968).
[174] *H. Hart*, *T. R. Rodgers*, and *J. Griffiths*, J. Amer. chem. Soc. *91*, 754 (1969).
[175] *R. F. Childs* and *S. Winstein*, J. Amer. chem. Soc. *90*, 7146 (1968).

Several examples of [3,4] cationic sigmatropic reactions have recently become available through the elegant researches of *Schmid*[176]. In these dienol-benzene, as well as the analogous dienone-phenol, rearrangements, the [3,4] sigmatropic reactions are competitive with equally allowed [1,2] and [3,3] shifts.

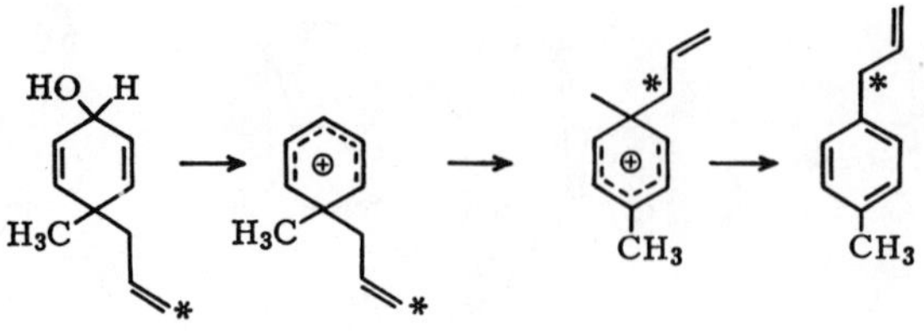

(363)

There also may be at hand some examples of [2,3] anionic sigmatropic reactions, in the Wittig rearrangement — the base-catalyzed transformation of benzyl ethers to alcohols. Thus, *Makisumi* and *Notzumoto*[177] observed the base-catalyzed conversion of *(364)* to *(365)*. Similar results on labeled 9-fluorenylallyl ethers were ob-

(364) *(365)*

tained by *Schöllkopf*[178]. The isoelectronic analogue [*(366)*→*(367)*], within formally neutral species, has been found to occur readily, and has been recognized as a very general reaction-type[179].

(366) *(367)*

[176] *H.-J. Hansen, B. Sutter*, and *H. Schmid*, Helv. chim. Acta *51*, 828 (1968).
[177] *Y. Makisumi* and *S. Notzumoto*, Tetrahedron Letters *1967*, 6393.
[178] *U. Schöllkopf* and *K. Fellenberger*, Liebigs Ann. Chem. *698*, 80 (1966). See also *H. E. Zimmerman* in *P. de Mayo*: Molecular Rearrangements. Interscience, New York 1963, Vol. 1, p. 372.
[179] *J. E. Baldwin, R. E. Hackler*, and *D. P. Kelly*, Chem. Commun. *1968*, 537, 538; *G. M. Blackburn, W. D. Ollis, J. D. Plackett, C. Smith*, and *I. O. Sutherland*, ibid. *1968*, 186; *R. B. Bates* and *D. Feld*, Tetrahedron Letters *1968*, 417; *B. M. Trost* and *R. LaRochelle*, ibid. *1968*, 3327; *J. E. Baldwin* and *R. E. Peavy*, ibid. *1968*, 5029; *W. Kirmse* and *M. Kapps*, Chem. Ber. *101*, 994, 1004 (1968).

We would expect that [1,2] anionic shifts should proceed with inversion at the migrating atom. Such shifts have been observed[180] in all-carbon systems, but stereochemical information is not available. The Stevens rearrangement involves the migration of an alkyl group from a quaternary ammonium atom to an adjacent carbanionoid center. The reaction is thus isoelectronic with a [1,2] carbanion rearrangement. It is definitely intramolecular[181], and migration proceeds with retention of configuration at the shifting carbon atom[182]. There is considerable argument concerning the mechanism of the reaction[180, 182, 183]. Ion-pair and diradical mechanisms have been suggested[182, 183] which might circumvent the disagreement with the predictions for a concerted process.

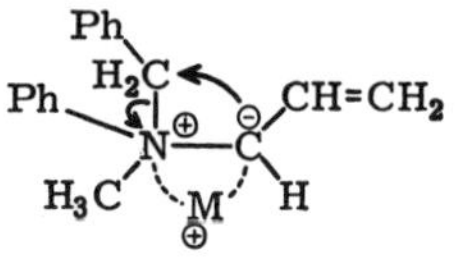

(368)

The perhaps bizarre alternative suggestion may be put forward that the metal ions always associated with such systems play a determinative role. If a metal ion is strongly associated with the carbanionoid center at the outset of reaction, and with the nitrogen atom when the migration has been concluded, the symmetry-allowed process depicted in *(368)* may be envisaged; in this change, inversion occurs at nitrogen and at the carbanionoid carbon, and a *p* orbital of the metal is involved.

---

[180] *E. Grovenstein Jr.* and *G. Wentworth*, J. Amer. chem. Soc. *89*, 1852, 2348 (1967), and references therein.

[181] *T. S. Stevens*, J. chem. Soc. *1930*, 2107; *R. A. W. Johnstone* and *T. S. Stevens*, ibid. *1955*, 4487; *R. K. Hill* and *T.-H. Chan*, J. Amer. chem. Soc. *88*, 866 (1966).

[182] *J. H. Brewster* and *M. W. Kline*, J. Amer. chem. Soc. *74*, 5179 (1952); *B. J. Millard* and *T. S. Stevens*, J. chem. Soc. *1963*, 3397; *E. F. Jenny* and *J. Druey*, Angew. Chem. *74*, 152 (1962); Angew. Chem. internat. Edit. *1*, 155 (1962).

[183] *D. J. Cram:* Fundamentals of Carbanion Chemistry. Academic Press, New York 1965, p. 223; *P. T. Lansbury*, *V. A. Pattison*, *J. D. Sidler*, and *J. B. Bieber*, J. Amer. chem. Soc. *88*, 78 (1966); *A. R. Lepley*, ibid. *91*, 1237 (1969); *J. E. Baldwin*, personal communication.

(369)

The homo [1,5] sigmatropic shift *(369)* is very well known[184].

7.2. Sequential Sigmatropic Shifts

The impetus for a general consideration of sequential sigmatropic shifts arose from the experimental observation by *Berson* and *Willcott* of a sequence of [1,5] sigmatropic shifts *(370)* in norcaradienes[161]. Such processes must proceed *with retention*

(370)

at the migrating carbon atom. The consequences of a sequence of motions with retention are sketched in *(371)*. The illustration is essentially a superposition of

---

[184] *D. S. Glass*, *J. Zirner*, and *S. Winstein*, Proc. chem. Soc. *1963*, 276; *R. J. Ellis* and *H. M. Frey*, ibid. *1964*, 221; J. chem. Soc. *1964*, Suppl. 1, 5578; *W. R. Roth*, Liebigs Ann. Chem. *671*, 10 (1964); *W. R. Roth* and *J. König*, ibid. *688*, 28 (1965); *W. Grimme*, Chem. Ber. *98*, 756 (1965); *J. K. Crandall* and *R. J. Watkins*, Tetrahedron Letters *1967*, 1717; *R. M. Roberts*, *R. G. Landolt*, *R. N. Greene*, and *E. W. Heyer*, J. Amer. chem. Soc. *89*, 1404 (1967); *G. Ohloff*, Tetrahedron Letters *1965*, 3795; *W. R. Roth*, Chimia *20*, 229 (1966). – The reaction is also known in the case of cyclopropyl-carbonyl compounds: *D. E. McGreer*, *N. W. K. Chiu*, and *R. S. McDaniel*, Proc. chem. Soc. *1964*, 415. A special case of some interest is the "abnormal Claisen rearrangement" studied by *E. N. Marvell*, *D. R. Anderson*, and *J. Ong*, J. org. Chem. *27*, 1109 (1962); *W. M. Lauer* and *T. A. Johnson*, ibid. *28*, 2913 (1963); *A. Habich*, *R. Barner*, *W. von Philipsborn*, and *H. Schmid*, Helv. chim. Acta *48*, 1297 (1964); *R. M. Roberts* and *R. G. Landolt*, J. Amer. chem. Soc. *87*, 2281 (1965); *R. M. Roberts*, *R. N. Greene*, *R. G. Landolt*, and *E. W. Heyer*, ibid. *87*, 2282 (1965); *R. M. Roberts*, *R. G. Landolt*, *R. N. Greene*, and *E. W. Heyer*, ibid. *89*, 1404 (1967). See also *J. M. Conia*, *F. Leyendecker*, and *C. Dubois-Faget*, Tetrahedron Letters *1966*, 129; *F. Rouessac* and *J. M. Conia*, ibid. *1965*, 3313.

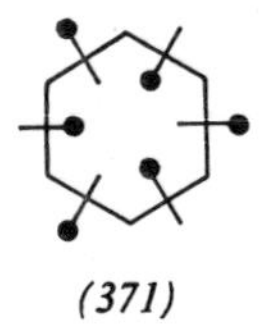

(371)

snapshots of all of the possible positions of labels attached to the methylene carbon atom as it migrates around the six-membered ring (*cf.* Figure 36).

Now it is also conceivable, though energetically unlikely, that the methylene group might migrate by a sequence of [1,3] sigmatropic shifts, through norbornadiene intermediates *(372)*.

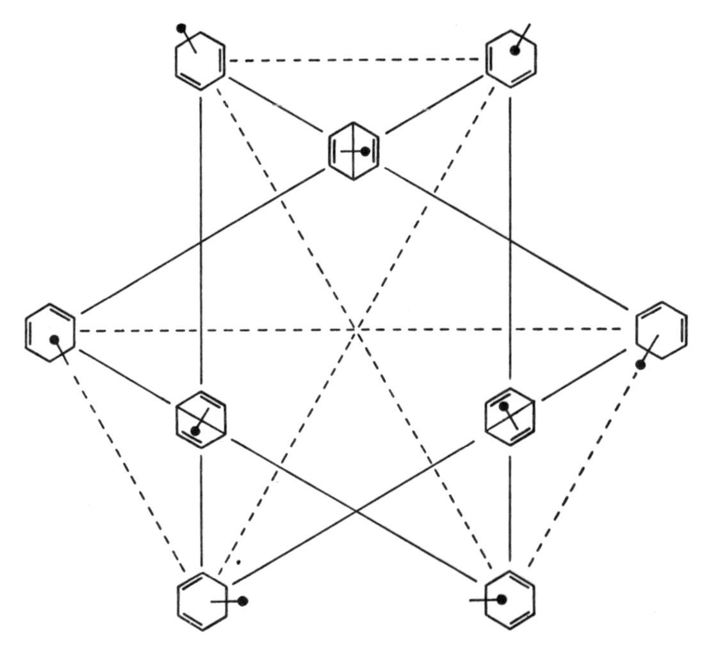

Figure 37. Sigmatropic [1,3] and [1,5] shifts in substituted norcaradienes and norbornadienes. — [1,3] shifts, ——— [1,5] shifts.

In that connection the following question may be asked: given that suprafacial [1,5] sigmatropic shifts must proceed with retention, and that [1,3] shifts must proceed with inversion if constrained to be suprafacial — is it true that a random

(372)

sequence of [1,3] and [1,5] shifts will produce only the same labeling pattern as the previously analyzed sequence of [1,5] shifts [*cf.* *(371)*]? In this case the answer is yes, as shown by a diagram (Figure 37) relating all the possible species through all the possible paths. The information contained in Figure 37 is summarized in *(373)* and *(374)*.

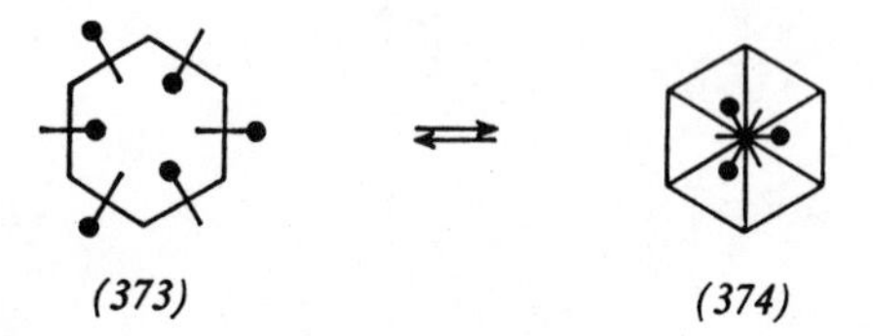

(373) *(374)*

Similar questions may be posed for other [2n.1.0] bicyclic systems. For the [2.1.0] system, only [1,3] shifts are possible; the labeling pattern is summarized in *(375)*. For the [6.1.0] system [1,3], [1,5], and [1,7] shifts are all possible. The diagram re-

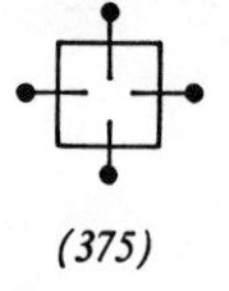

(375)

lating all isomers is very complicated, but its information content may be summarized in *(376)* and *(377)*. Here again any sequence of [1,3], [1,5], or [1,7] shifts which relates two isomers gives the same configuration as any other sequence.

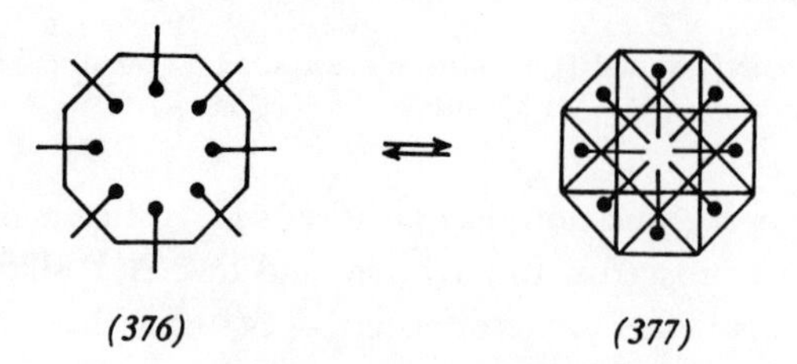

(376) *(377)*

It may be shown that the results derived above represent a general topological property of [2n.1.0] bicyclic systems.

There are two parts to the proof. First, it is clear that a [1,5] shift with retention at a migrating center is indistinguishable from a sequence of *two* [1,3] shifts, each proceeding with inversion at the migrating center. Similarly, a [1,7] shift with inversion is stereochemically indistinguishable from a series of *three* [1,3] shifts, each with inversion. In general,

$$[1,(2q+1)] = [1,3]^q$$

Consequently, it is sufficient to consider a random sequence of [1,3] shifts in constructing a proof.

We will prove that in a bicyclo[(2n-2).1.0] system *every* possible sequence of [1,3] shifts which effects the transformation of *(378)* into *(379)* contains an odd num-

(378) *(379)*

ber of such shifts if *n* is even, and an even number of shifts if *n* is odd. Thus, it will be shown that for *n* even, there is an odd number of inversions, *i. e.* one net inversion [*(380)*→*(381)*], whereas for *n* odd, there is net retention [*(382)*→*(383)*]. The cyclic symmetry of the rearrangement then generates the isomer relationships shown in *(384)*.

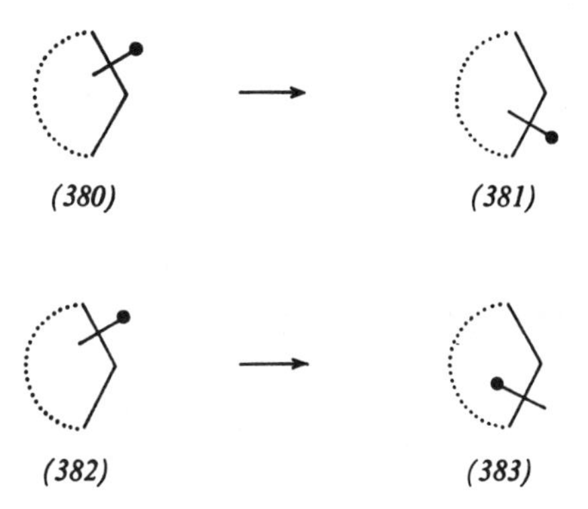

(380) *(381)*

(382) *(383)*

We now construct the mathematical problem equivalent to the rearrangement sequence. Consider a polygon *(385)* with $2n$ vertices, numbered sequentially clock-

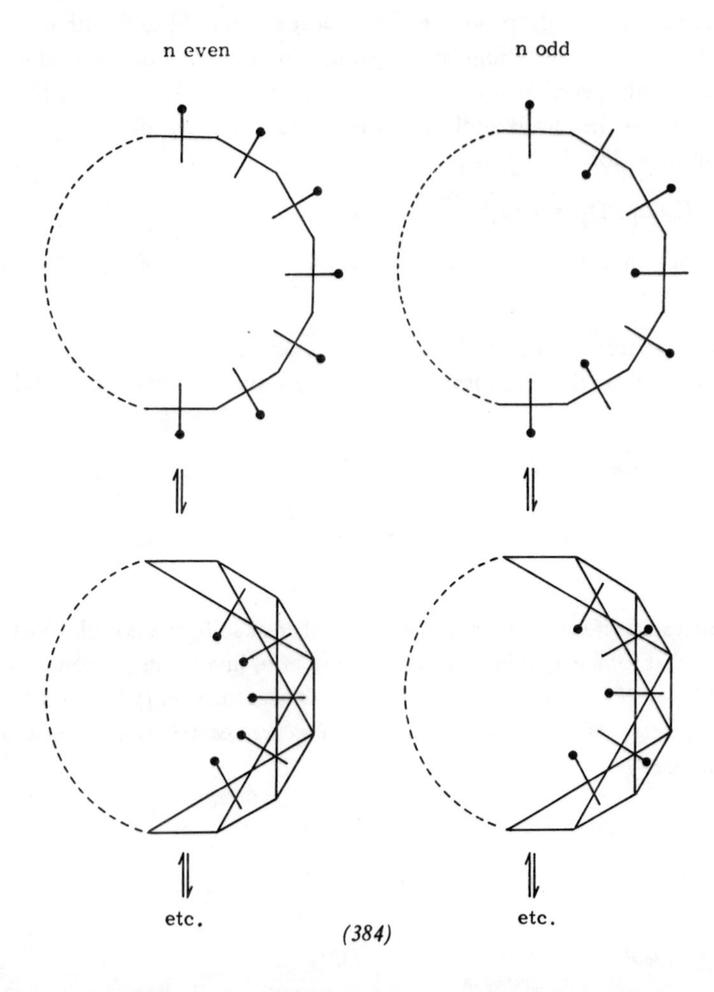

(384)

wise from an arbitrary origin. Suppose markers, A and B, are placed at vertices 1 and 2, respectively, of this polygon. A game is then played, with the following set of rules:

a) A random number of moves of the markers to new vertices is performed sequentially.

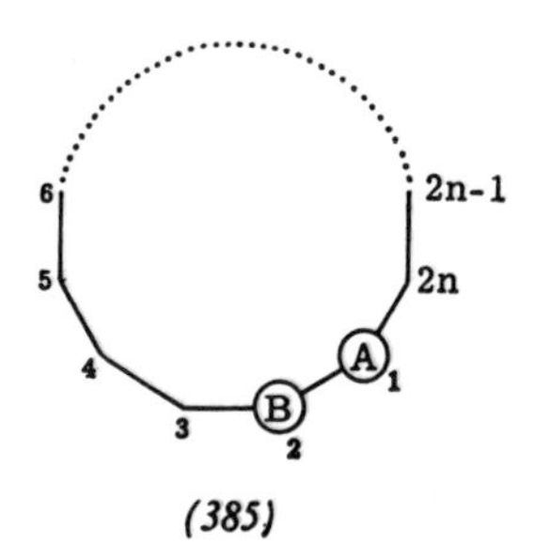

(385)

b) In each move *either* A *or* B moves two positions clockwise or counterclockwise. Thus, if A is at position k, it can move to $k+2$ or $k-2$.

c) No hopping is allowed; *i.e.* if A is located at point k and B at $k-1$, the move of A from k to $k-2$ is forbidden and A has only the option of moving to $k+2$.

Theorem: Given the above rules of the game, *every possible sequence of moves* which begins with A at 1 and B at 2 and terminates with A at 3 and B at 2 comprises an even number of moves if n is odd, and an odd number of moves if n is even.

Before proceeding to the proof of this theorem, one should assure oneself that the above game and theorem are equivalent to the problem presented by the sequential rearrangements. The requisite identifications are that the ring of $2n$ atoms, around which the carbon atom moves, corresponds to the polygon, while the markers A and B represent the bonds from the vertices of the polygon to the migrating atom.

Proof: Define a distance, measured in number of sides of the polygon, between the markers A and B, beginning at B, and *measured clockwise*. Then if B is at 2 and A is at 1, the distance between the two is the maximum: $2n-1$. We now proceed to examine the number of steps it takes to reduce this distance between the markers to 1, with B at 2 and A at 3. First, we will examine the number of moves it takes to reduce the distance to 1, *when B may have moved to any vertex*. The initial move reduces the distance from $2n-1$ to $2n-3$. Every subsequent move can be classified as *effectual* or *ineffectual*. An effectual move makes the distance smaller than the previous minimum, an ineffectual move does not. All ineffectual moves come in pairs: for, if at some point the distance is j, and a move increases the distance to $j+2$, then another move is necessary to return it to j. Only effectual moves reduce the distance between A and B. Since each effectual move reduces the distance by 2, the total number of moves, for any sequence, necessary to reduce the distance from $2n-1$ to 1 is:

$$\underset{\textit{Effectual}}{(2n-2)/2 = (n-1)} + \underset{\textit{Ineffectual}}{\text{some even number}}$$

So far we have deduced the number of moves necessary to reduce the distance to 1, but B may be at any even-numbered vertex. Now, A and B having been placed at *any* adjacent vertices, an even number of ineffectual moves will suffice to realize the final objective of placing B at position 2, and A at position 3. We have thus shown that the total number of necessary moves is:

$$(n-1) + \text{some even number}$$

Consequently, the required number of moves is odd if n is even, and even if n is odd, which was to be proven.

For cationic sigmatropic rearrangements, the topological relationships can differ significantly from those just shown to be general for neutral species. We consider first the sigmatropic rearrangements of a bicyclo[3.1.0]hexenyl cation *(386)*. In

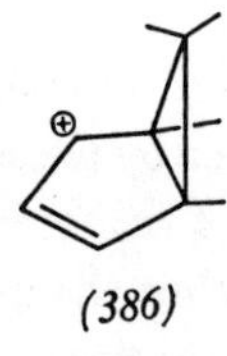

(386)

this system [1,2], [1,4], and [1,3] shifts may occur; experience indicates that ionic [1,2q] shifts should be much faster than the neutral [1,(2q+1)] rearrangements. Using the previous conventions, and observing the symmetry-imposed conditions that [1,2] shifts must occur with retention, while [1,3] and [1,4] migrations must take place with inversion, we obtain the interrelations shown in Figure 38. The relationships are summarized in *(387)* and *(388)*.

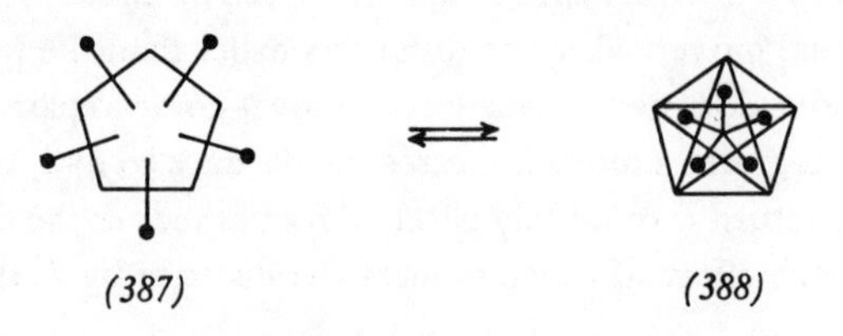

(387) *(388)*

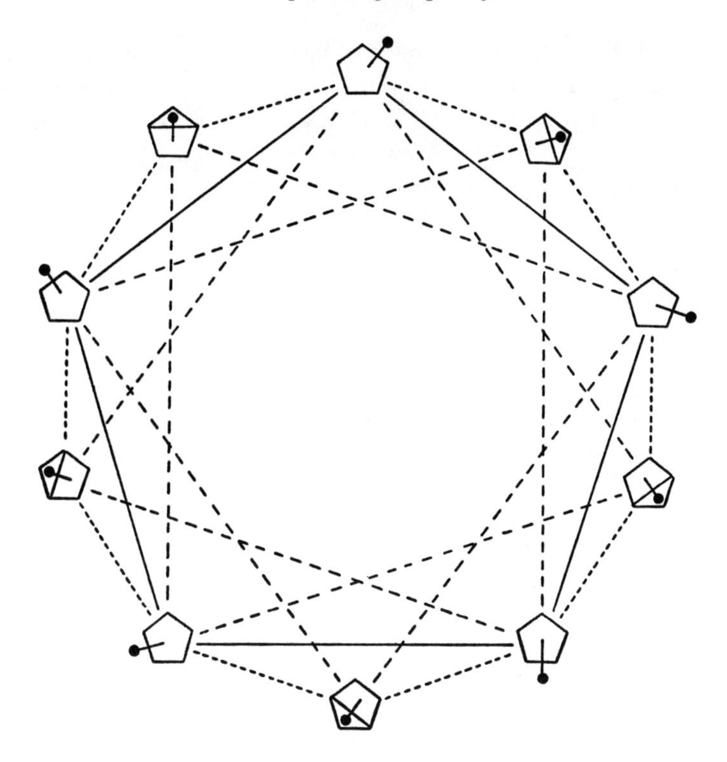

Figure 38. Sigmatropic [1,4], [1,2], and [1,3] shifts in the cation *(386)*. — [1,4] shifts, ---- [1,2] shifts, - - - - - [1,3] shifts.

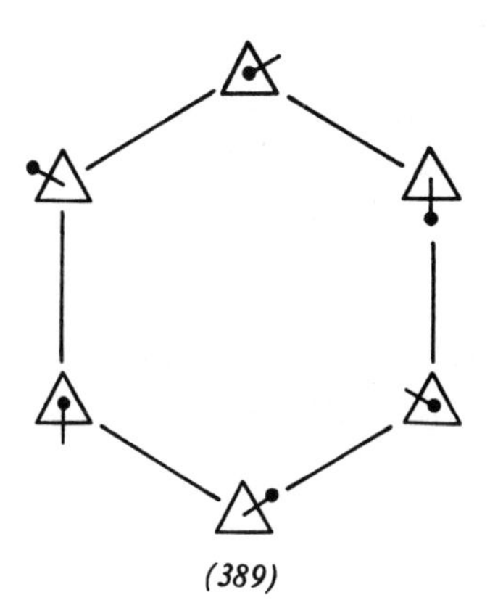

(389)

At this point it might appear that sequential sigmatropic reactions in cyclic polyenyl ions will always have the same consequences as in cyclic polyenes. *This is not so.* Consider the simplest system in which [1,2] shifts are possible — the rearrangements of the bicyclo[1.1.0]butyl cation *(389)*. In sharp contrast to those in each of the previous cases studied, this sequence produces apparent epimerization, *i.e.* a series of rearrangements can convert *(390)* into *(391)*. It may be readily seen that

(390) *(391)*

similar circumstances will obtain in (7, 11, . . .)-membered rings — as well as in the three-membered case (*cf.* Figure 39).

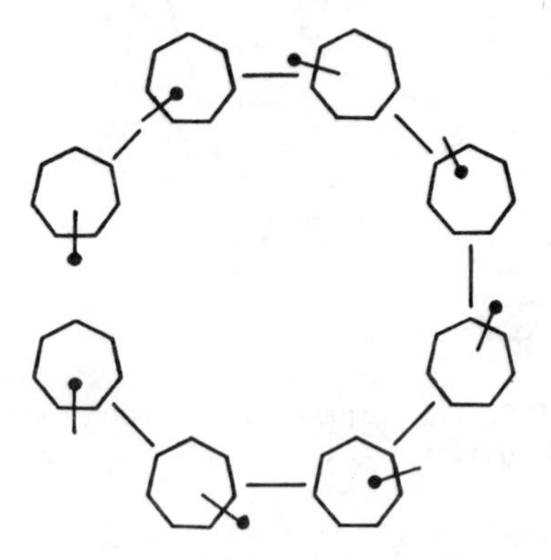

Figure 39. [1,6] Sigmatropic shifts in bicyclo[5.1.0]octadienyl cations.

The bicyclo[(2n+1).1.0] cation cases thus must be divided into two categories. In the first class, the [1,(2n+2)] sigmatropic shift proceeds with *retention* and epimerization must occur as a consequence of $2n+3$ such shifts; such is the case for [1,2], [1,6], [1,10], . . . shifts, that is, in general for cases where n is even. The second category comprises those cases in which n is odd, the [1,(2n+2)] sigmatropic shift proceeds with *inversion* and epimerization can never be realized through any number of symmetry-allowed migrations — as in the cyclic polyene cases.

8. Group Transfers and Eliminations

Consider the concerted reaction in which two hydrogen atoms are transferred from ethane to ethylene *(392)*, in a process suprafacial on both reactants. A correlation

(392)

diagram (Figure 40) is easily constructed for this reaction, utilizing the symmetry plane bisecting the two molecules; the process is clearly symmetry-allowed. In contrast is the diagram (Figure 41) for the concerted transfer of two hydrogen

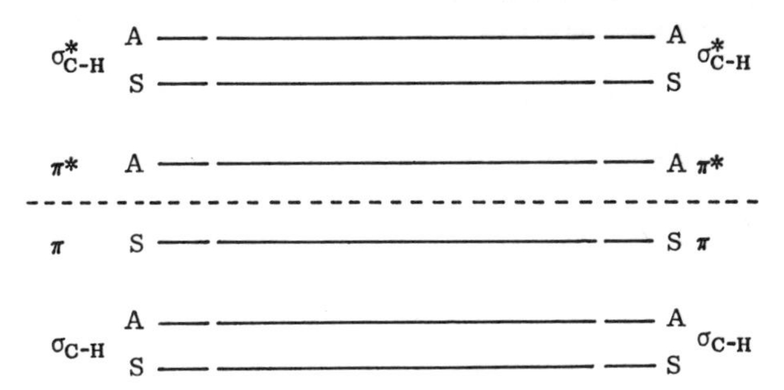

Figure 40. Correlation diagram for the concerted transfer of two hydrogen atoms from ethane to ethylene.

atoms from ethane to the termini of butadiene *(393)*; the reaction is symmetry-forbidden.

(393)

For the general case *(394)*, it is simple to derive the following selection rule: The double-group transfer is symmetry-allowed for ground states when $m+n = 4q+2$, for excited states when $m+n = 4q$, where m and n are numbers of π electrons, and q

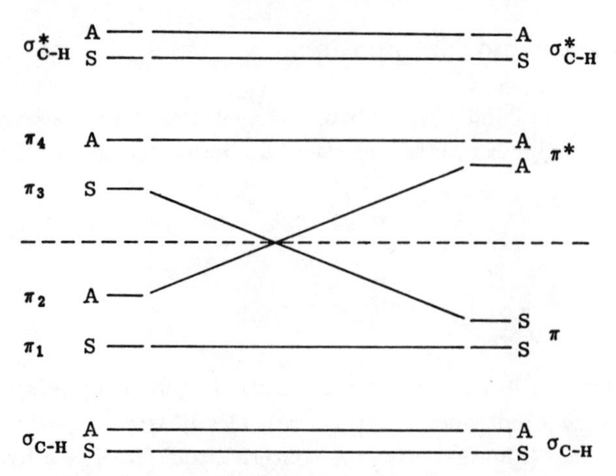

Figure 41. Correlation diagram for the concerted transfer of two hydrogen atoms from ethane to the termini of butadiene.

is an integer (0, 1, 2, . . .). The rule applies also to a process antarafacial on both components, and is reversed for a process antarafacial on one component only.

m R R n → m+2 R R n-2

(394)

The rules must be appropriately modified if the possibility of inversion at R is real (*cf.* Section 11).

The best known example of this type of reaction, with $m = 0, n = 2$, is the transfer of hydrogen from diimide to olefins[185]. Hydrocarbon analogues also exist; for exam-

H H + H_3C H_3C → + H_3C H H_3C H

(395)

[185] Reviewed by *S. Hünig, H. R. Müller*, and *W. Thier*, Angew. Chem. *77*, 368 (1965); Angew. Chem. internat. Edit. *4*, 271 (1965); *E. J. Corey, D. J. Pasto*, and *W. L. Mock*, J. Amer. chem. Soc. *83*, 2957 (1961).

ple, the reaction *(395)* was observed by *Doering* and *Rosenthal*[186]. A case with $m = 2, n = 4$ has also been observed *(396)*[187].

(396)

The rule derived above also applies to the limiting case with $n = 0$ — the concerted elimination *(397)*. We predicted, thus, that the concerted noncatalytic hydrogenation of a diene, or the reverse elimination, should be 1,4 rather than 1,2. At the

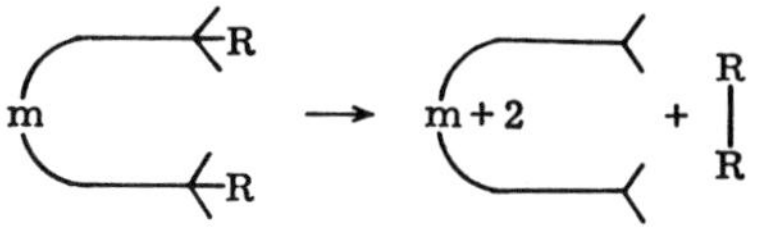

(397)

time our predictions were made the literature appeared to contain only two relevant pieces of information:

a) 2,5-Dihydrofuran loses hydrogen in a unimolecular process with $E_a = 48.5$ kcal/mole and a negative entropy of activation, while 2,3-dihydrofuran does not react in this way[188].

b) 3,3,6,6-Tetramethyl-1,4-cyclohexadiene when pyrolyzed in pentane gave a much larger yield of ethane (and *p*-xylene) than could have arisen from a free radical chain reaction[189].

Subsequently, our predictions have been substantiated in a number of studies. *Baldwin*[190] in studying the elimination of hydrogen from an isotopically labeled cyclopentene found a greater than 10:1 preference for 1,4 over 1,2 elimination. *El-*

[186] *W. von E. Doering* and *J. W. Rosenthal*, J. Amer. chem. Soc. *89*, 4535 (1967).
[187] *P. Dowd*, personal communication; *I. Fleming*, personal communication.
[188] *C. A. Wellington* and *W. D. Walters*, J. Amer. chem. Soc. *83*, 4888 (1961).
[189] *W. Reusch*, *M. Russell*, and *C. Dzurella*, J. org. Chem. *29*, 2446 (1964).
[190] *J. E. Baldwin*, Tetrahedron Letters *1966*, 2953.

lis and *Frey*[191], *Benson* and *Shaw*[192], and *Kellner* and co-workers[193] have independently studied the pyrolysis of 1,3 and of 1,4-cyclohexadiene. The latter decomposes unimolecularly to benzene and hydrogen with $E_a = 42.7$ kcal/mole; at the same temperatures the 1,3-cyclohexadiene is thermally stable and at higher temperatures it decomposes mainly by radical-chain processes. *Frey* has provided further interesting examples of concerted elimination of hydrogen; on the other hand, his evidence indicates that those processes which lead to the formation of methane or ethane proceed by radical-chain mechanisms[194].

The concerted loss of two geminal groups *(398)*, provided that it proceeds through a transition state of C_{2v} symmetry (*cf.* Section 10.1), should be an excited state reaction, and thus compete with vicinal elimination [*(397)*, $m = 0$] — also a symmetry-allowed excited state process. Both cases are known[195].

$$CR_2 \rightarrow C: + R{-}R$$

(398)

[191] *R. J. Ellis* and *H. M. Frey*, J. chem. Soc. *A*, *1966*, 553.

[192] *S. W. Benson* and *R. Shaw*, Trans. Faraday Soc. *63*, 985 (1967); J. Amer. chem. Soc. *89*, 5351 (1967).

[193] *W. D. Walters*, personal communication.

[194] *H. M. Frey* and *D. H. Lister*, J. chem. Soc. *A*, *1967*, 509, 1800; *H. M. Frey* and *R. Walsh*, Chem. Rev. *69*, 103 (1969).

[195] Photolysis of cyclohexadiene: *R. Srinivasan*, J. Amer. chem. Soc. *83*, 2806 (1961); photolysis of ethane and ethylene: *M. Okabe* and *J. R. McNesby*, J. chem. Physics *34*, 668 (1962); cyclohexane: *R. P. Doepker* and *P. Ausloos*, ibid. *42*, 3746 (1965). In the gas phase photolysis of azomethane, it is well established that the predominant primary process is decomposition to nitrogen and two methyl radicals; however, some evidence has been presented for a minor molecular elimination of ethane: *R. E. Rebbert* and *P. Ausloos*, J. phys. Chem. *66*, 2253 (1962).

9. Secondary Effects

Orbital symmetry arguments may also be used to yield further insight into the origins of secondary conformational effects in concerted cycloaddition reactions.

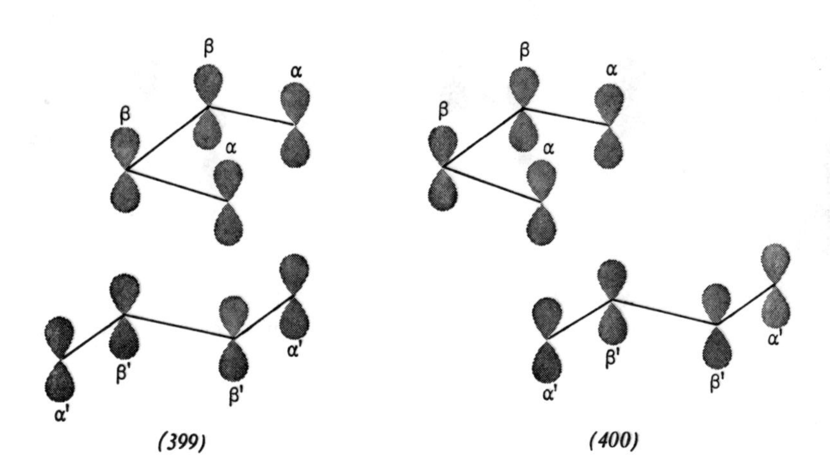

(399) *(400)*

Consider the Diels-Alder addition of butadiene to itself, as a concerted $[_{\pi}4_s+_{\pi}2_s]$ cycloaddition reaction. *A priori*, this reaction might take place through either of two alternative transition states, *(399)* or *(400)*. The *endo* approach *(399)* is distinguished from the *exo* *(400)* mainly by the proximity of a β and a β' orbital in the former. Any secondary interaction among occupied diene and dienophile molecular orbitals will contribute only in a very minor way to the total energy of the transition state, since such an interaction will increase the energy of some levels while decreasing the energy of others. The significant interactions will come from the symmetry-allowed mixing of unoccupied with occupied levels[196]. In the case at hand, inspection of diagrams *(401)* and *(402)* reveals at once that either possibility for such mixing leads to bonding, *i.e.* energy-lowering, interaction of the proximate β and β' orbitals[197]. Thus, the *endo* transition state *(399)* for this $[_{\pi}4_s+_{\pi}2_s]$

---

[196] This has also been stressed by *K. Fukui* in *P.-O. Löwdin* and *B. Pullman:* Molecular Orbitals in Chemistry, Physics, and Biology. Academic Press, New York 1964, p. 525.
[197] The role of other secondary interactions has been stressed by *L. Salem*, J. Amer. chem. Soc. *90*, 543, 553 (1968).

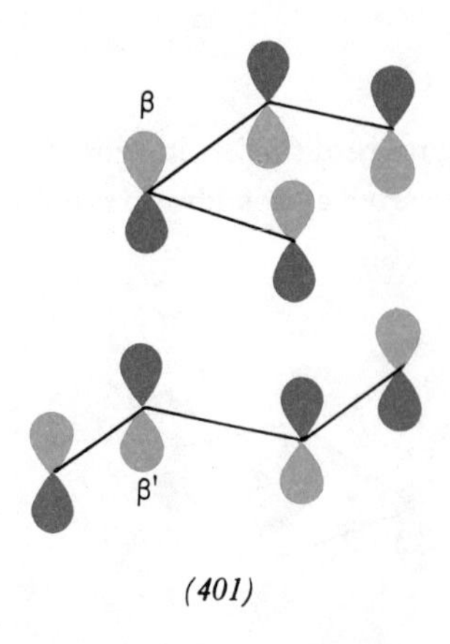

(401)

Mixing of highest occupied "diene" orbital (top) with lowest unoccupied "olefin" orbital (bottom)

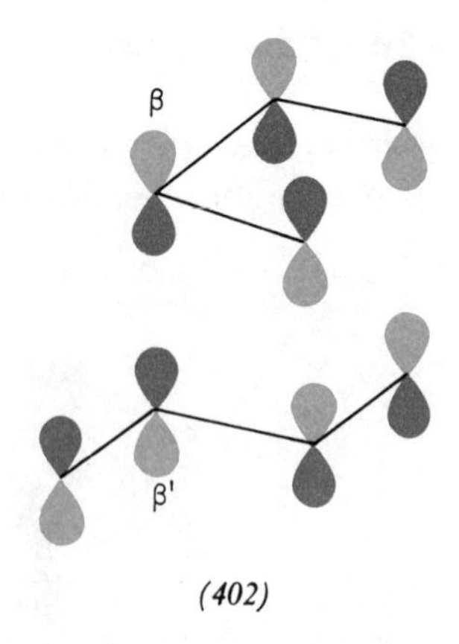

(402)

Mixing of lowest unoccupied "diene" orbital (top) with highest occupied "olefin" orbital (bottum)

concerted cycloaddition reaction is stabilized *vis-a-vis* the *exo* alternative *(400)* by symmetry-controlled secondary orbital interactions.

The orbital symmetry relationships signalized here provide a simple quantum chemical basis for the large body of experience summarized in the Alder *endo* addition rule[198]. Our treatment differs from previous proposals, which have emphasized the roles of inductive forces[199], electrostatic forces consequent upon charge transfer between diene and dienophile[200], and maximum accumulation of unsaturation[198]. In particular, it is now clear that in some cases the orbital interactions among unsaturated centers involved in a concerted cycloaddition reaction will be such as to raise, rather than lower, the energy of the *endo* transition state, and lead to a preference for *exo* addition, insofar as symmetry factors are dominant; inspection of the relevant orbital diagrams *(403)* and *(404)* for the symmetry-allowed $[_{\pi}6_s + _{\pi}4_s]$ combinations indicates that it is such a case.

---

[198] *K. Alder* and *G. Stein*, Angew. Chem. *50*, 514 (1937); *K. Alder*, Liebigs Ann. Chem. *571*, 157 (1951); *K. Alder* and *M. Schumacher*, Fortschr. Chem. org. Naturstoffe *10*, 1 (1953). For exceptions to the rule, *cf. J. A. Berson, Z. Hamlet*, and *W. A. Mueller*, J. Amer. chem. Soc. *84*, 297 (1962).

[199] *A. Wassermann*, J. chem. Soc. *1935*, 825, 1511; *1936*, 432; Trans. Faraday Soc. *35*, 841 (1939).

[200] *R. B. Woodward* and *H. Baer*, J. Amer. chem. Soc. *66*, 645 (1944).

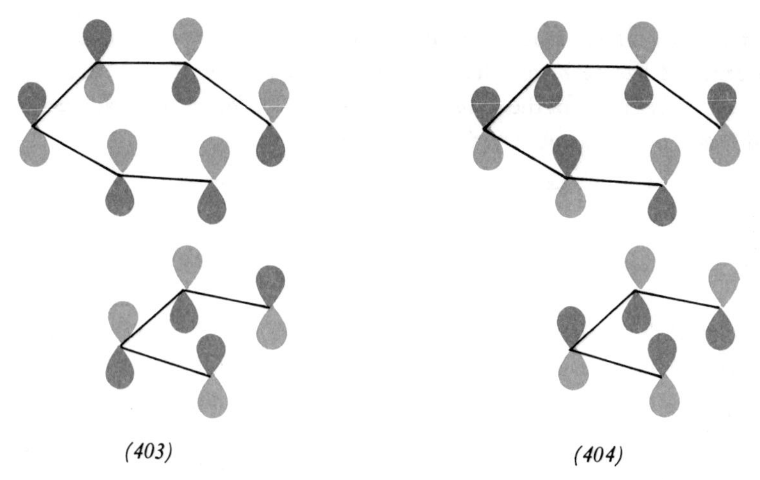

(403) (404)

This prediction was made prior to the discovery of [6+4] cycloadditions, and has since been confirmed[201]. By contrast, the symmetry-allowed $[_{\pi}8_s+_{\pi}2_s]$ and $[_{\pi}2_s+_{\pi}2_s+_{\pi}2_s]$ cycloadditions should resemble the [4+2] process and proceed by preference through *endo* transition states; there is already evidence for that preference in an example of the latter process[202].

The dimerization of cyclobutadiene is a special case of much interest, considered in the light of orbital symmetry relationships. If, as seems likely, there is substantial double bond localization in that highly reactive molecule, one can delineate, *a priori*, the possibilities of $[_{\pi}2_s+_{\pi}2_s]$, $[_{\pi}2_s+_{\pi}4_s]$, and $[_{\pi}4_s+_{\pi}4_s]$ cycloadditions. Our selection rules require a preference for the concerted $[_{\pi}2_s+_{\pi}4_s]$ process, and indeed evidence that this path is favored has been brought forward[203]. Examination of secondary orbital interactions along the lines set down here reveals further that the *endo* process, leading to the *syn* dimer *(405)*, should be favored over the alternative *exo* combination, which would give the *anti* dimer *(406)*. The formation of both

(405) (406)

[201] *R. C. Cookson, B. V. Drake, J. Hudec,* and *A. Morrison,* Chem. Commun. *1966,* 15; *K. Houk,* Dissertation, Harvard (1968).

[202] *R. C. Cookson, J. Dance,* and *J. Hudec,* J. chem. Soc. *1964,* 5416.

[203] *G. Wittig* and *J. Weinlich,* Chem. Ber. *98,* 471 (1965).

syn and *anti* dimers in reactions initiated with a variety of possible cyclobutadiene precursors has been reported, but in those experiments in which there was the greatest likelihood of the transitory existence of a free cyclobutadiene, the predicted *endo* process was observed[204].

The [3,3] sigmatropic shift in 1,5-hexadienes has been shown to proceed more easily through a four-center chair-like transition state *(407)* than through the boat-like alternative *(408)*[205]. We suggest that orbital symmetry relationships may play

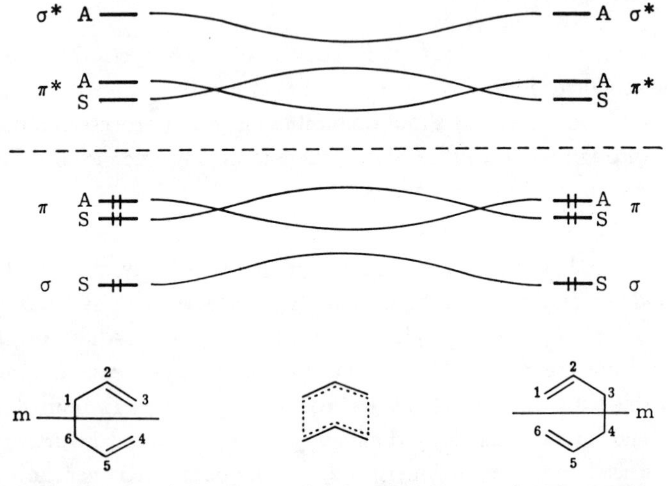

Figure 42. Correlation diagram for the [3,3] sigmatropic rearrangement of 1,5-hexadiene.

---

[204] *R. Criegee*, Angew. Chem. *74*, 703 (1962); Angew. Chem. internat. Edit. *1*, 519 (1962), and references therein; *P. S. Skell* and *R. J. Peterson*, J. Amer. chem. Soc. *86*, 2530 (1964), and references therein; *L. Watts*, *J. D. Fitzpatrick*, and *R. Pettit*, ibid. *88*, 623 (1966); *E. Hedaya*, *R. D. Miller*, *D. W. McNeil*, *P. F. D' Angelo*, and *P. Schissel*, J. Amer. chem. Soc. *91*, 1875 (1969).

[205] *W. von E. Doering* and *W. R. Roth*, Tetrahedron *18*, 67 (1962); Angew. Chem. *75*, 27 (1963); Angew. Chem. internat. Edit. *2*, 115 (1963); *R. K. Hill* and *N. W. Gilman*, Chem. Commun. *1967*, 619; for some recent theoretical work see *M. Simonetta*, *G. Favini*, *C. Mariani*, and *P. Gramaccioni*, J. Amer. chem. Soc. *90*, 1280 (1968). Similar preferences in the aromatic Claisen rearrangement have been elegantly studied by *Gy. Fráter*, *A. Habich*, *H.-J. Hansen*, and *H. Schmid*, Helv. chim. Acta *52*, 335 (1969).

a major role in determining that preference. A correlation diagram for the molecular orbitals involved in the rearrangement is illustrated in Figure 42. The levels are classified as symmetric or antisymmetric with respect to the mirror plane *m* in the boat-like transition state, or a two-fold rotation axis in the chair-like form. The scheme shown is for the former case, and the diagram for the latter is qualitatively similar. The correlation of reactant bonding levels with product bonding levels, characteristic of a symmetry-allowed thermal reaction, should be noted. At the half-way mark in the reaction the level ordering is recognizable as that of two strongly interacting allyl radicals. The actual behavior of the levels along the reaction coordinate is abstracted from extended Hückel calculations.

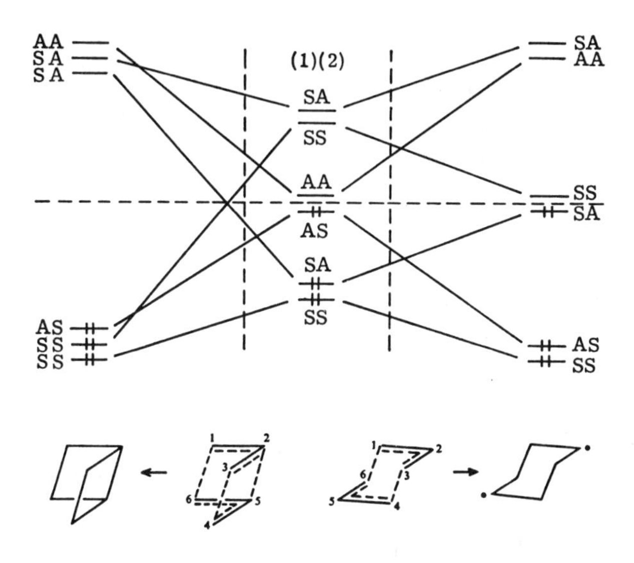

Figure 43. Correlation diagram for the approach of two allyl radicals from infinity in a *quasi* boat or a *quasi* chair relationship. (1) and (2) at the top of the center column refer to the symmetry elements present (see text).

If the correlation diagrams are qualitatively similar for the boat and chair reactions, where does the observed preference come from? An explanation requires the construction of two further correlation diagrams (Figure 43), for the alternative hypothetical processes in which two allyl radicals approach each other from infinity, in parallel planes, oriented with respect to each other in the one case in a *quasi* boat relationship, and in the other in a *quasi* chair fashion. In each of these processes there are two symmetry elements:

a) in both, a mirror plane m_1, which passes through C-2 and C-5, and bisects the angles C(1)-C(2)-C(3) and C(6)-C(5)-C(4);

b) in the boat approach, a second mirror plane, m_2, parallel to and half-way between the planes of the approaching radicals, and in the chair approach a two-fold axis of symmetry, lying on the intersection of m_2 and the plane defined by C(1)-C(3)-C(4)-C(6).

To complete the correlations one must specify the end products of these hypothetical motions; these are a bicyclohexane in the boat approach, and a 1,4-cyclohexylidene biradical in the chair pathway. Among the occupied levels the critical difference in the two pathways is in the behavior of the occupied S_1A_2 level which in the boat-like approach correlates to an antibonding σ orbital while in the chair-like process it goes over to a non-bonding biradical level. The crucial point of the argument now is that a reaction proceeding as in Figure 42 must pass at the half-way point through some point in a correlation diagram resembling one or the other of those shown in Figure 43, and that this point will be approximately the same horizontal distance, marked by dashed vertical lines, along the reaction coordinate in the two alternate pathways of Figure 43. Further, at any such point the chair-like transition state is at lower energy as a result of the difference in correlation properties of the occupied S_1A_2 orbital. The central conclusion is that orbital interactions involving C-2 and C-5 have a net antibonding, energy-raising effect. In this sense, the argument represents a further development of the simple orbital repulsion effect suggested by *Doering* and *Berry*[205].

It should be emphasized that the effects discussed here are, not unexpectedly, small ones, and that in systems possessing special geometrical restraints which necessitate a boat-like transition state, [3,3] sigmatropic changes take place with no special difficulty[206].

To present a final illustration of the procedure, in a case in which the result is not yet known, we consider the [4+2] cycloaddition of a butadiene to an allyl cation.

---

[206] *J. M. Brown*, Proc. chem. Soc. *1965*, 226; *W. von E. Doering* and *W. R. Roth*, Tetrahedron *19*, 715 (1963); *R. Merényi, J. F. M. Oth*, and *G. Schröder*, Chem. Ber. *97*, 3150 (1964); *H. A. Staab* and *F. Vögtle*, Tetrahedron Letters *1965*, 54.

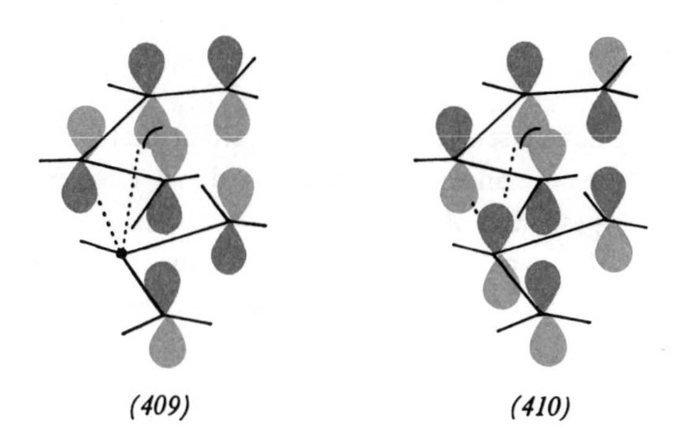

(409) *(410)*

The two significant interactions are shown in *(409)* and *(410)*. The secondary interactions in *(409)* must be negligible, while those in *(410)* are antibonding; a preference for the *exo* transition state is expected[206a].

[206a] Since the above was written, the prediction has been confirmed: *H. M. R. Hoffmann* and *D. R. Joy*, J. chem. Soc. *B*, *1968*, 1182.

10. Divertissements!

In the previous sections, we have discussed and exemplified orbital symmetry control of the major general types of concerted organic reactions. We turn now to a discussion of two special types, each of which possesses fascinating unique features, and further illustrates the power of the principle of orbital symmetry conservation in understanding and predicting the detailed course of chemical transformations.

10.1. Cheletropic Reactions

We define as *cheletropic reactions* those processes in which two σ bonds which terminate at a single atom are made, or broken, in concert.

C.1
m-2 X (YZ...) → m + X (YZ...)
C.m

(411)

Let us consider the cheletropic reaction *(411)*, in which a small molecule X(YZ. . .) and a polyene containing an *m*-electron π system are produced. We shall examine the geometrical aspects of the reaction, with reference to an invariant coordinate system *(412)*, whose origin is always at X, and whose z axis bisects the line between C. 1 and C. m. Thus, the σ bonds being broken lie in the xz plane, and

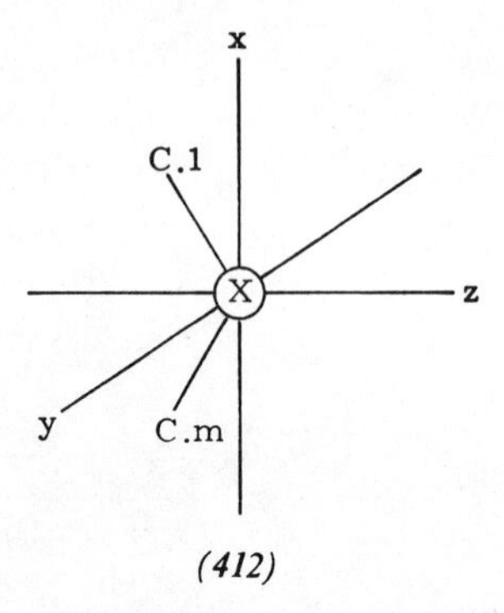

(412)

their four electrons are distributed as usual, in pairs, in the two generalized orbitals *(413)* and *(414)*. Now, in general, two electrons must be delivered from these orbitals to each of the developing products, with conservation of orbital symmetry. A crucial point now emerges: the transfer of two electrons from the breaking σ bonds to new orbitals in the product molecule X(YZ. . .) may take place in either of two sharply distinct ways.

a) Two electrons from the symmetric orbital [*(413)*$\equiv$*(415)*] may take up new positions in a z-symmetric lone pair orbital of X(YZ. . .), or enter a π system antisymmetric with respect to the xy plane.

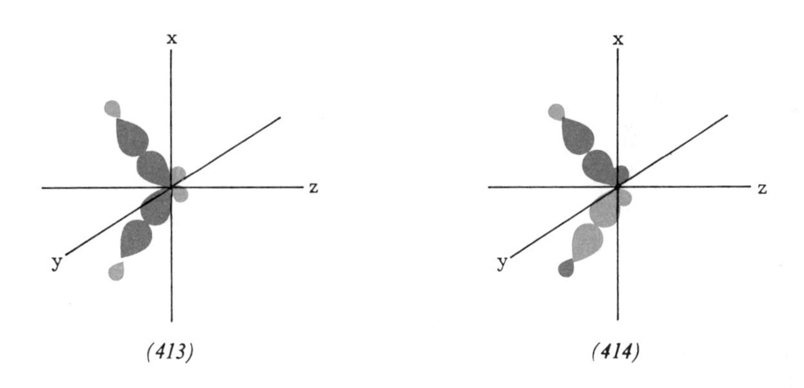

(413) *(414)*

b) Two electrons from the antisymmetric orbital [*(414)*$\equiv$*(416)*] may take up new positions in an x-symmetric lone pair orbital of X(YZ. . .), or enter a π system antisymmetric with respect to the yz plane.

In the light of this analysis, it is clear that the geometrical displacements at C. 1 and C. m are dependent upon the detailed geometry of departure of X(YZ. . .), and in particular, on the sense of the displacements of the atoms YZ. . . which are attached to X. To illustrate this important point, we consider the departure of a bent molecule Y-X-Z, which as it develops must acquire a symmetric lone pair at the central atom X. The reaction may take place in either of two ways, depicted in *(417)* and *(418)*.

a) *Cf. (417)*: X acquires its needed lone pair from the symmetric σ orbital *(415)*, as X, Y, and Z depart co-linearly, all remaining in the yz plane. We shall designate such processes as *linear* cheletropic reactions. It may be noted that in a formal sense, the product electron pair is participating in a *suprafacial* fashion, in that the two bonds from which the two electrons came lie on the same side of the nodal

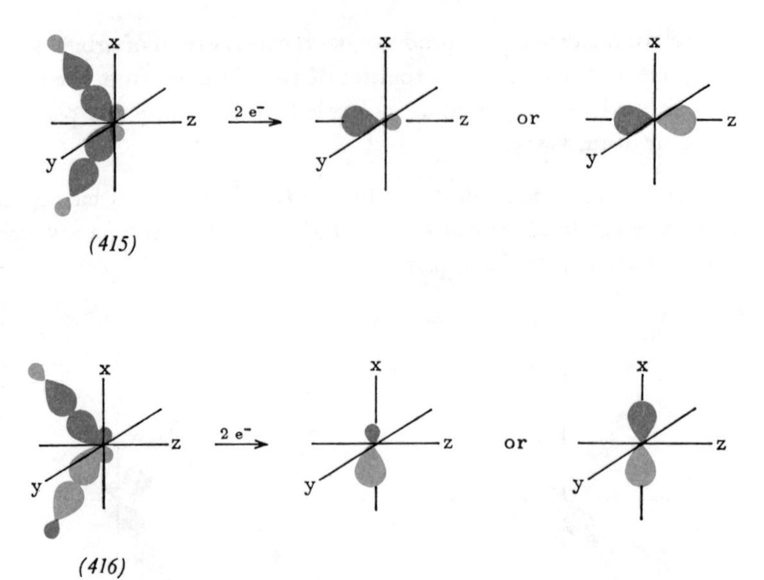

(415)

(416)

surface (approximated by the plane xy) of the product orbital which the pair occupies.

b) *Cf. (418)*: X acquires its needed lone pair from the antisymmetric σ orbital *(416)*, as X departs co-linearly, remaining in the yz plane, while Y and Z undergo displacements in both the x and z directions, with the result that the product molecule lies in the xy plane. We shall designate such processes as *non-linear* cheletropic

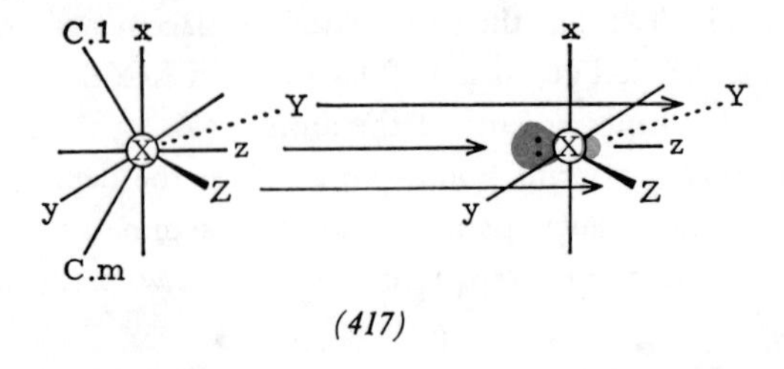

(417)

reactions. In this case the product electron pair at X is participating in an *antarafacial* fashion, in that the two bonds from which the electrons came lie on opposite sides of the nodal surface (approximated by the plane yz) of the product orbital which the pair occupies.

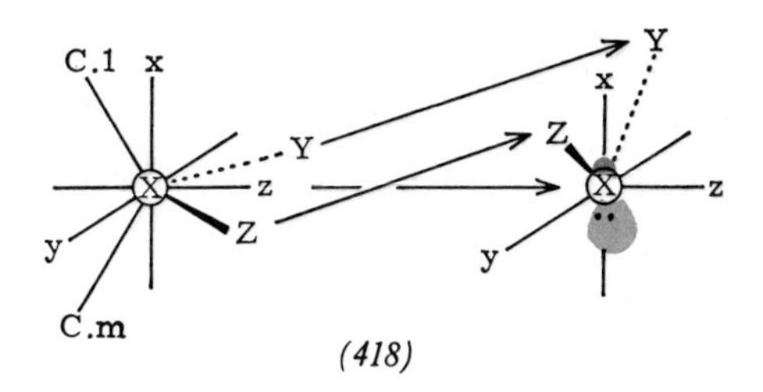

(418)

It is now clear that in linear cheletropic reactions of the type *(411)*, two electrons must be delivered from the antisymmetric σ orbital [*(416)* ≡ *(420)* ≡ *(423)*] to the highest occupied π orbital of the polyene *(419)*. Consequently, the displacements at C.1 and C.m must be disrotatory when $m = 4q$ [*(420)*→*(421)*], and conrotatory when $m = 4q+2$ [*(423)*→*(424)*].

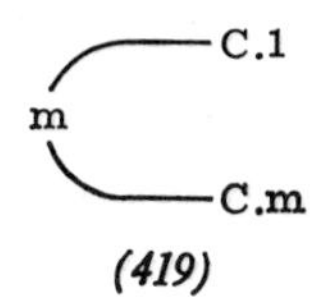

(419)

By contrast, if the fragmentation *(411)* follows the non-linear cheletropic path, two electrons must be delivered to the highest occupied π orbital of the polyene *(419)* from the symmetric σ orbital [*(415)* ≡ *(422)* ≡ *(425)*]. In this case, therefore, the displacements at C.1 and C.m must be conrotatory if $m = 4q$ [*(422)*→*(421)*], and disrotatory when $m = 4q+2$ [*(425)*→*(424)*].

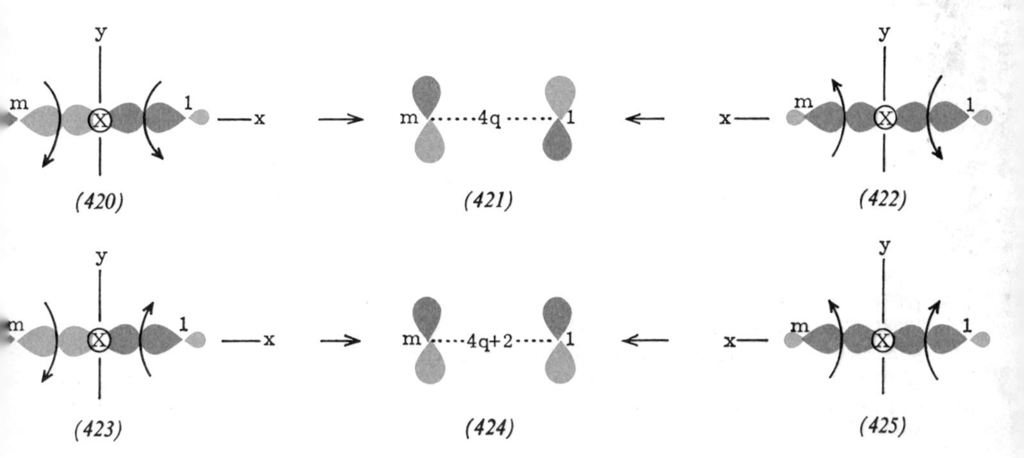

(420) (421) (422)

(423) (424) (425)

m	Allowed Ground State Reactions	
	Linear	Non-linear
$4q$	Disrotatory	Conrotatory
$4q+2$	Conrotatory	Disrotatory

Figure 44. Selection rules for cheletropic reactions.

The selection rules are summarized in Figure 44. In these cheletropic transformations there undoubtedly exists, for each ground state reaction, an excited state process characterized by reversed stereochemical features. However, the extruded small molecule X(YZ)... will frequently possess many occupied lone pair orbitals, and the accompanying possibility of numerous $n \rightarrow \pi^*$ states raises the possibility in any given case that excited state processes with the *same* stereochemical characteristics as the corresponding ground state transformation may be available.

Linear cheletropic fragmentations leading to the production of nitrogen or carbon monoxide represent reactions well-suited for the construction of formal correlation diagrams, which have been drawn by *Lemal*[207], and by *Baldwin*[208], as well as by us. For carbon monoxide elimination no direct stereochemical results are available. It is clear, however, that cyclopent-3-en-1-ones ($m = 4$) undergo decarbonylation readily, apparently in a concerted manner[208]. In this connection, the failure of attempts to prepare norbornadien-7-one *(426)*, may be contrasted with the stability of norbornanones[209].

O → + CO

(426)

A case with $m = 6$ is represented by 3,5-cycloheptadienone *(427)*, which in one of the few known condensed-phase decarbonylations, undergoes photolysis to hexatriene and carbon monoxide[210].

[207] *D. M. Lemal* and *S. D. McGregor*, J. Amer. chem. Soc. *88*, 1335 (1966).
[208] *J. E. Baldwin*, Canad. J. Chem. *44*, 2051 (1966).
[209] See, for example, *S. Yankelevich* and *B. Fuchs*, Tetrahedron Letters *1967*, 4945.
[210] *O. L. Chapman* and *G. W. Borden*, J. org. Chem. *26*, 4185 (1961); *O. L. Chapman*, *D. J. Pasto*, *G. W. Borden*, and *A. A. Griswold*, J. Amer. chem. Soc. *84*, 1220 (1962); *D. I. Schuster*, *B. R. Sckolnick*, and *F.-T. H. Lee*, ibid. *90*, 1300 (1968).

The elimination of nitrogen from diazenes *(428)* has been carefully studied by *Lemal*[207]. The process is stereospecific and disrotatory, as predicted; the linear disrotatory path is clearly one involving less strain than the non-linear conrotatory alternative in this case.

(427) *(428)*

Remarkable sequences [*(429)*, *(430)*] comprising disrotatory nitrogen eliminations, followed by conrotatory electrocyclic reactions have been revealed by *Carpino*[211].

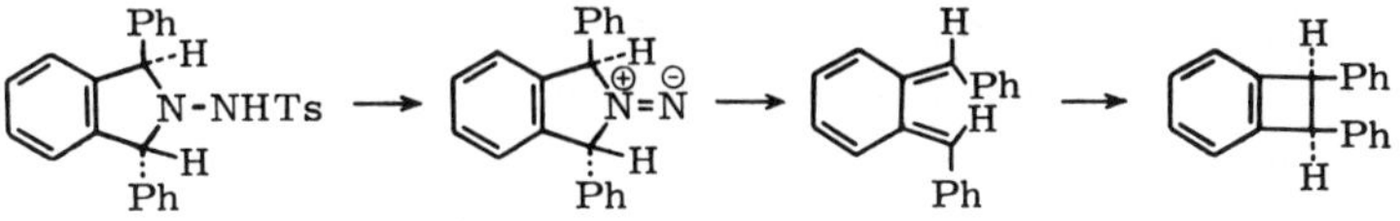

(429)

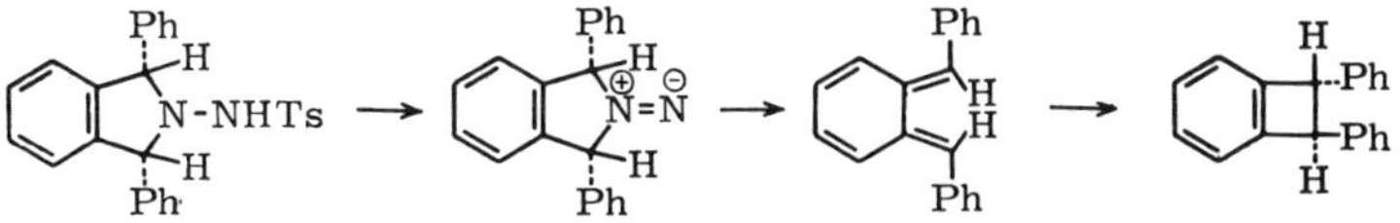

(430)

The reversible, facile addition of sulfur dioxide[212] to dienes is well known[213]. The expected disrotatory course for the elimination of sulfur dioxide from a butadiene sulfone has been established[214]. The corresponding photochemical process is

[211] *L. A. Carpino*, Chem. Commun. *1966*, 494.

[212] For molecular orbital descriptions of sulfur dioxide and sulfones, see *W. Moffitt*, Proc. Roy. Soc. *A200*, 409 (1950).

[213] *H. Staudinger* and *B. Ritzenhaler*, Ber. dtsch. chem. Ges. *68*, 455 (1935); *G. Hesse*, *E. Reichold*, and *S. Majmudar*, Chem. Ber. *90*, 2106 (1957); *M. P. Cava* and *A. A. Deana*, J. Amer. chem. Soc. *81*, 4266 (1959). The last group has also described some interesting selective photochemical eliminations of sulfur dioxide: *M. P. Cava*, *R. H. Schlessinger*, and *J. P. Van Meter*, J. Amer. chem. Soc. *86*, 3173 (1964).

[214] *W. L. Mock*, J. Amer. chem. Soc. *88*, 2857 (1966); *S. D. McGregor* and *D. M. Lemal*, ibid. *88*, 2858 (1966).

not completely stereospecific, but a clear preference for the conrotatory path is shown[215]. Sulfur dioxide also readily undergoes 1,6-addition to 3-*cis*-hexatrienes; the reverse process has recently been shown to proceed in a conrotatory fashion, and must therefore be classified as a linear cheletropic reaction[216].

Cheletropic reactions in which a three-membered ring is produced or destroyed, *e.g.* fragmentations of type *(411)* in which $m = 2$, are of special interest in that geometric factors require that they take place in a disrotatory, or suprafacial manner. Consequently, these transformations must occur by the non-linear cheletropic path.

The well-known[217] stereospecific combination of singlet carbenes with olefins to give cyclopropanes undoubtedly falls in the non-linear cheletropic class[218].

Cyclopropenones and cyclopropanones lose carbon monoxide thermally under mild conditions[219, 220], and even more readily on irradiation[219, 221]. Stereochemical information on the decarbonylation is lacking. It has, however, been clearly demonstrated that nitrogen is stereospecifically extruded in a disrotatory fashion from alkyl-substituted three-membered ring diazenes *(431)*[222]; this change must follow the non-linear path, with x-displacement of N-2 in the plane of the aziridine ring as reaction proceeds.

---

[215] *J. Saltiel* and *L. Metts*, J. Amer. chem. Soc. *89*, 2232 (1967).
[216] *W. L. Mock*, J. Amer. chem. Soc. *89*, 1281 (1967); *W. L. Mock*, ibid. *91*, 5682 (1969).
[217] *P. S. Skell* and *R. C. Woodworth*, J. Amer. chem. Soc. *78*, 4496 (1956); *81*, 3383 (1959); *P. S. Skell* and *A. Y. Garner*, ibid. *78*, 5430 (1956); *W. von E. Doering* and *P. LaFlamme*, ibid. *78*, 5447 (1956); *H. M. Frey*, ibid. *80*, 5005 (1958); *W. R. Moore*, *W. R. Moser*, and *J. E. LaPrade*, J. org. Chem. *28*, 2200 (1963).
[218] For this reaction, semiempirical calculations yield results in striking concordance with the analysis of non-linear cheletropic processes presented here. *Cf. R. Hoffmann*, J. Amer. chem. Soc. *90*, 1475 (1968).
[219] *R. Breslow*, *T. Eicher*, *A. Krebs*, *R. A. Peterson*, and *J. Posner*, J. Amer. chem. Soc. *87*, 1320 (1965); *R. Breslow*, *L. J. Altman*, *A. Krebs*, *E. Mohacsi*, *I. Murata*, *R. A. Peterson*, and *J. Posner*, ibid. *87*, 1326 (1965); *R. Breslow* and *G. Ryan*, ibid. *89*, 3073 (1967).
[220] *N. J. Turro*, *P. A. Leermakers*, *H. R. Wilson*, *D. C. Neckers*, *G. W. Byers*, and *G. F. Vesley*, J. Amer. chem. Soc. *87*, 2613 (1965).
[221] *D. C. Zecher* and *R. West*, J. Amer. chem. Soc. *89*, 153 (1967).
[222] *J. P. Freeman* and *W. H. Graham*, J. Amer. chem. Soc. *89*, 1761 (1967). However, the corresponding aryl-substituted case proceeds in a non-stereospecific manner (personal communication from *L. A. Carpino*).

(431)

Sulfur dioxide is readily lost from episulfones in a stereospecific disrotatory manner[223, 224]. In order to circumvent the problem which seemed to be posed by the fact that linear cheletropic fragmentation is symmetry-forbidden in this case, it has been suggested[224] that the reaction might proceed in a non-concerted way, through a discrete intermediate. It is now clear that the available evidence is also consistent with the view that the elimination follows the concerted symmetry-allowed non-linear cheletropic path [*cf.* *(418)*].

The observed ready elimination of nitrous oxide *(433)* from *N*-nitrosoaziridines[225] probably also follows the non-linear cheletropic path [*(432)*→*(433)*];

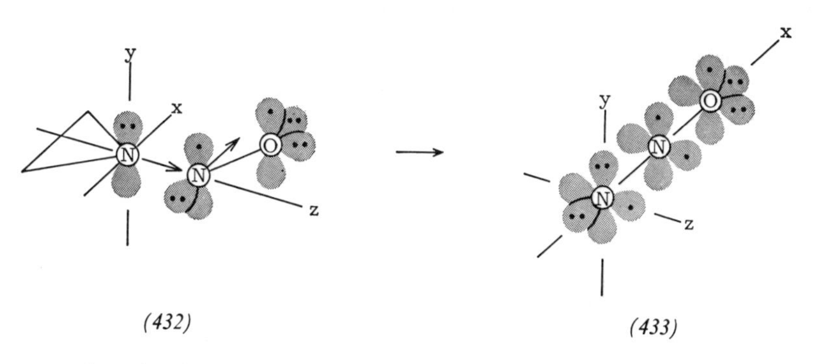

(432) *(433)*

note that the $4\pi_y$ system remains intact throughout the reaction, and that an x-symmetric lone pair at nitrogen in *(433)* is derived from the bonding antisymmetric σ orbital of the aziridine ring. However, an interesting alternative should be considered. If the nitroso compound be allowed to achieve the conformation *(434)*, the unshared electron pair at the pyramidal aziridine nitrogen atom may become one of the z-symmetric lone pairs of nitrous oxide *(435)*, while the electron

---

[223] *N. P. Neureiter* and *F. G. Bordwell*, J. Amer. chem. Soc. *85*, 1210 (1963); *N. Tokura, T. Nagai*, and *S. Matsumara*, J. org. Chem. *31*, 349 (1966); *L. A. Carpino* and *L. V. McAdams III*, J. Amer. chem. Soc. *87*, 5804 (1965); *N. P. Neureiter*, ibid. *88*, 558 (1966); *L. A. Paquette*, Accounts Chem. Res. *1*, 209 (1968).

[224] *F. G. Bordwell, J. M. Williams, Jr., E. B. Hoyt, Jr.*, and *B. B. Jarvis*, J. Amer. chem. Soc. *90*, 429 (1968).

[225] *R. D. Clark* and *G. K. Helmkamp*, J. org. Chem. *29*, 1316 (1964).

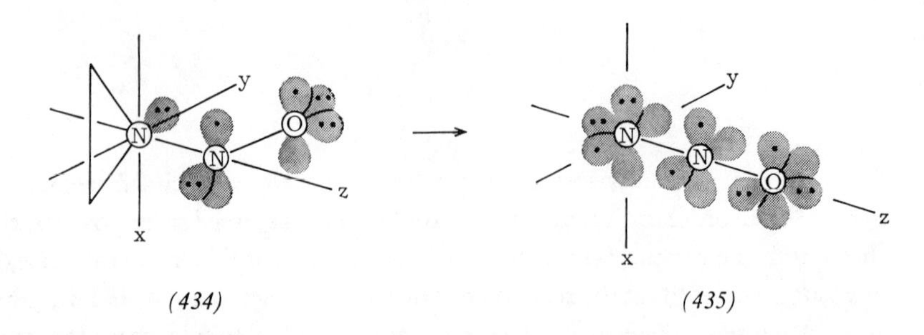

(434) (435)

pair from the bonding antisymmetric σ orbital of the aziridine ring enters the $4\pi_x$ system. In view of the fact that the uncoupling of two electrons from the $4\pi_y$ system of *(432)* may require *ca.* 23 kcal/mole, as estimated from the barrier to rotation in *N*-nitrosoamines[226], it seems unlikely that the molecule will make use of this alternative symmetry-allowed process to destroy itself.

The hypothetical abstraction reaction *(436)* provides a further example of a cheletropic change. If the abstraction takes place in a *linear* fashion, the lone pair of the

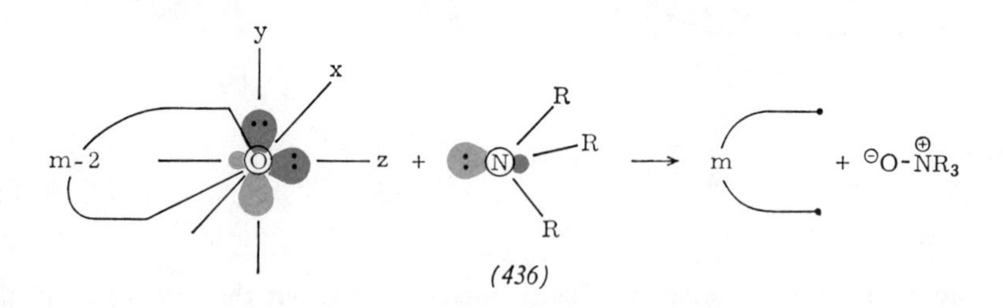

(436)

amine and the two oxygen lone pairs are transformed into the N-O bond and the z and y-axis unshared pairs of the *N*-oxide *(437)*. The remaining x-axis lone pair of the *N*-oxide must be taken from the antisymmetric σ orbital of the breaking carbon-oxygen bonds. Therefore, two electrons from the corresponding symmetric σ orbital must be delivered to the polyene product. Consequently, the reaction must occur in a disrotatory sense when $m = 4q+2$, or conrotatory when $m = 4q$. If, on the other hand, the amine attacks in a *non-linear* manner, *i.e.* in the xz plane, but in the x direction, the stereochemical consequences are precisely reversed. A possible ana-

[226] *C E. Looney*, *W. D. Phillips*, and *E. L. Reilly*, J. Amer. chem. Soc. *79*, 6136 (1957).

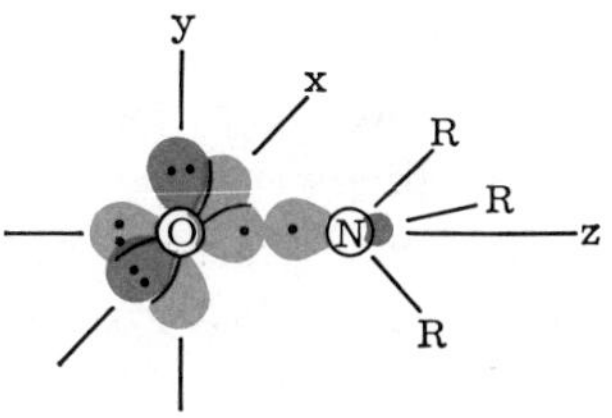

(437)

logue of the linear reaction is available in the stereospecific abstraction of sulfur from the *cis*- and *trans*-butene episulfides by phosphines[227].

A fascinating concatenation of a cheletropic fragmentation with a vicinal elimination is presented by the decomposition of a spiroketal to carbon dioxide and two olefins. As yet, the only known instances of the reaction involve ethylene ketals derived from substituted norbornadienones *(438)*[228]. It is most interesting to exam-

(438)

ine the stereochemical aspects of the process in the general case *(439)*; *m* and *n* represent the numbers of electrons in the π systems of the product polyenes. In addition to the four electrons which constitute the skeletal σ bonds, carbon dioxide

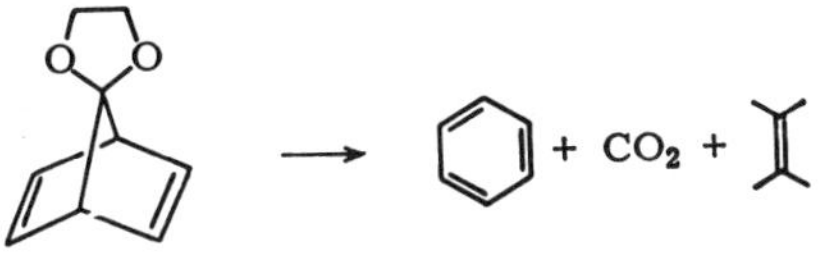

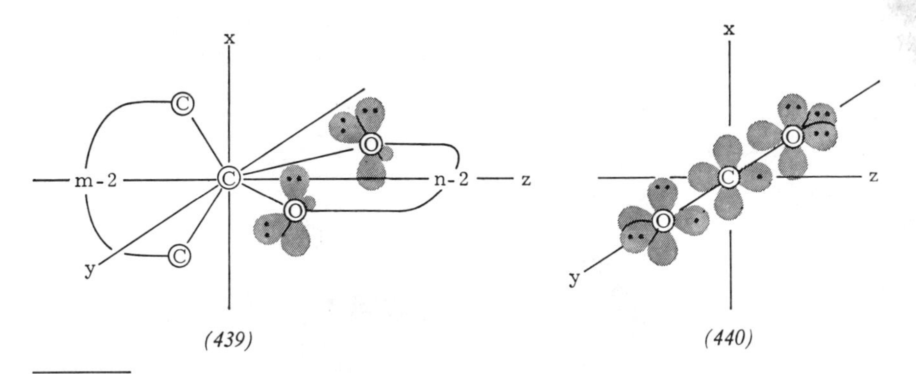

(439) *(440)*

[227] *D. B. Denney* and *M. J. Boskin*, J. Amer. chem. Soc. *82*, 4736 (1960).
[228] *D. M. Lemal, E. P. Gosselink*, and *S. D. McGregor*, J. Amer. chem. Soc. *88*, 528 (1966).

(440) requires twelve electrons: eight of these populate two perpendicular 4π systems, in the xy and yz planes, respectively, while four occupy two y-symmetric lone pair orbitals. If the reaction under discussion follows the linear cheletropic path, it is clear that the eight lone pair electrons at the oxygen atoms of *(439)* pass smoothly into the y-symmetric lone pair orbitals and the $4\pi_{xy}$ system of carbon dioxide *(440)*, with conservation of orbital symmetry. Further, the orbitals of the orthogonal $4\pi_{yz}$ system can only be filled by transferring two electrons from the symmetric C-C-C σ orbital *(441)* to the lowest occupied symmetric π_{yz} orbital *(442)*, and two from the antisymmetric O-C-O σ orbital *(444)* to the highest occupied π_{yz} orbital *(443)*. Therefore, the electrons occupying the antisymmetric

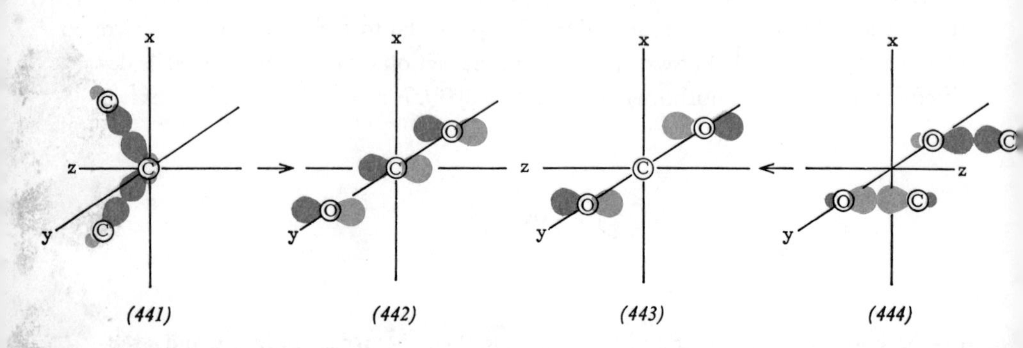

C-C-C σ orbital *(446)* must be delivered to the highest occupied orbital of the polyene *(445)*, while those in the symmetric O-C-O σ orbital *(447)* take up their new positions in the highest occupied orbital of the polyene *(448)*. Consequently, the

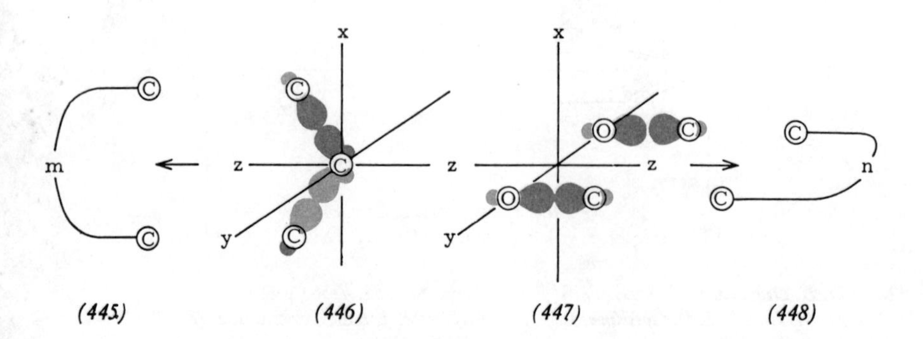

geometric displacements associated with the carbon-carbon bond cleavages will be disrotatory when $m = 4q$, or conrotatory when $m = 4q+2$; by contrast, the carbon-oxygen scissions will be associated with disrotatory displacements when $n = 4q+2$, or conrotatory when $n = 4q$. If the reaction follows the *non-linear* cheletropic path, the stereochemical consequences attendant upon the carbon-carbon bond cleavages will of course be reversed. Finally, it may be noted that if there is a minimum in the potential surface for the reaction, representing the intermediacy of a singlet carbene species *(449)*, the analysis in terms of orbital symmetry conservation is unchanged.

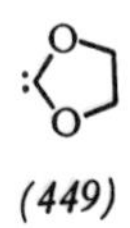

(449)

It might appear that in respect to cheletropic reactions, the principle of orbital symmetry conservation is rather generous than discriminate. But that is not so. Cheletropic reactions are unusual in that the stereochemical imperatives imposed by orbital symmetry control upon one of the reactant sites leave no imprint upon the products. Consequently, verification of the predictions which can be made about the detailed course of cheletropic reactions cannot be achieved simply by examining the stereochemistry of the products, and awaits more sophisticated scrutiny.

10.2. Cycloaddition Reactions of Ketenes

The combination of ketenes with olefins to give cyclobutanes has long been known[229]. In recent years, the reaction has been subjected to very searching scrutiny, and the evidence is now conclusive that it is concerted. Thus, the reaction proceeds with a high negative entropy of activation[230], the dependence of rate on solvent polarity is modest[230, 231], and stereometric relations in the reactants are maintained in the products[232] — the most striking instance being the formation of distinct adducts from the reactions of dichloroketene with the *cis*- and *trans*-cy-

[229] *Cf. H. Staudinger:* Die Ketene. Enke, Stuttgart 1912.

[230] *R. Huisgen, L. A. Feiler,* and *P. Otto,* Tetrahedron Letters *1968,* 4485.

[231] *W. T. Brady* and *H. R. O'Neal,* J. org. Chem. *32,* 612 (1967).

[232] *G. Binsch, L. A. Feiler,* and *R. Huisgen,* Tetrahedron Letters *1968,* 4497; *J. C. Martin, V. W. Goodlett,* and *R. D. Burpitt,* J. org. Chem. *30,* 4309 (1965); *R. Huisgen, L. Feiler,* and *G. Binsch,* Angew. Chem. *76,* 892 (1964); Angew. Chem. internat. Edit. *3,* 753 (1964).

clooctenes[233]. Most compelling is the observation that the reaction of ketenes with dienes — even those so constrained as to favor terminal addition — gives only cyclobutanes, and no cyclohexenes[234]. For example, cyclopentadiene and diphenylketene react to give *(450)*; no *(451)* is produced.

(450) *(451)*

A concerted combination of ethylenes to give a cyclobutane must follow the $[_\pi 2_s + _\pi 2_a]$ path, and in the case at hand the facts demonstrate beyond question that the olefinic reactant participates in a suprafacial way. Is there a special factor in the structure of ketenes which disposes them to play an antarafacial role in cycloaddition reactions?

We have seen that the $[_\pi 2_s + _\pi 2_a]$ process involves orthogonal approach of the reacting species (*cf.* Section 6) — *i.e.* that the π systems must orient themselves as in *(452)* — and we recognize that the path is not normally a readily accessible one, in

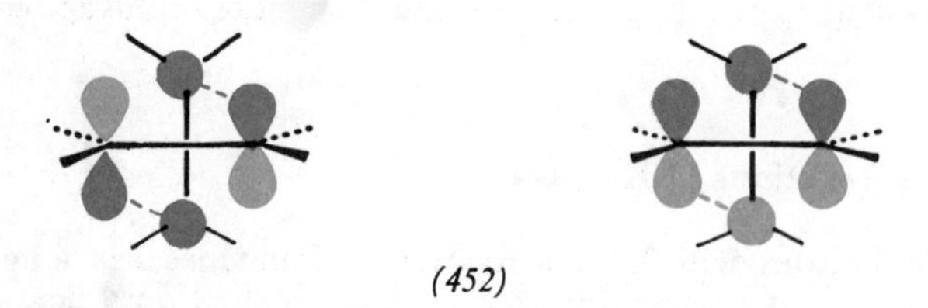

(452)

the absence of special factors (*cf.* Figure 25, Section 6.1). For normal olefins, insurmountable difficulties would appear to be presented by steric hindrance factors, and by transition-state strain associated with framework distortions necessary to maintain effective orbital overlap.

Ketenes are powerful electrophiles. Chemically, the spearhead of their reactivity is the carbon atom of their especially reactive carbonyl group. This dominant aspect

---

[233] *R. Montaigne* and *L. Ghosez*, Angew. Chem. *80*, 194 (1968); Angew. Chem. internat. Edit. *7*, 221 (1968).
[234] *R. Huisgen* and *P. Otto*, Tetrahedron Letters *1968*, 4491.

of the structure of ketenes, succinctly expressed in the valence-bond structure *(453)*, suggests that we analyze the combination of a vinyl cation *(454)* — *i.e.* a *vinylium* ion — with an olefin from the viewpoint of orbital symmetry conservation.

$$>C=\overset{\oplus}{C}-\overset{\ominus}{O} \qquad\qquad >C=\overset{\oplus}{C}-$$

(453) *(454)*

The relevant orbital diagrams are *(455)* and *(456)*, and the significant differences *vis-a-vis* those for the simple olefin-olefin combination *(452)* are the substantially more favorable situation in respect to steric oppositions, and more especially, the presence of the orthogonal vacant p orbital in the vinylium ion [green in *(455)* and checkered in *(456)*]. *It is at once apparent that the latter special feature contributes*

(455) *(456)*

two strong bonding interactions [dotted lines in *(455)*] *which are absent from the* $[_{\pi}2_s+_{\pi}2_a]$ *reaction path for simple olefins.* It should be noted that the view presented is such that the *top* of the vacant p orbital is seen, and that it is the bottom of that orbital which is bonding to the occupied π system of the simple olefinic reactant [*cf.* *(455)*]. To put the matter in another way, the normal symmetry-allowed combination of a cation with an ethylene, in this unique situation sets the stage for and is coalescent with the $[_{\pi}2_s+_{\pi}2_a]$ cycloaddition reaction.

In the light of the analysis just presented, it is clear that a vinylium ion must be regarded as constituted *par excellence* for participation as the $_{\pi}2_a$ component in the concerted $[_{\pi}2_s+_{\pi}2_a]$ cycloaddition. Consequently, it is most gratifying to find that the predicted reaction has in fact been observed, though not hitherto recognized as such. Thus, *Griesbaum*[235] found that allene reacts with hydrogen chloride at —70°C to give a mixture of the stereoisomeric dichlorodimethylcyclobutanes *(457)*; it is now clear that an initially-produced cation *(458)* combines with a molecule of al-

---

[235] *K. Griesbaum*, *W. Naegele*, and *G. G. Wanless*, J. Amer. chem. Soc. *87*, 3151 (1965); *K. Griesbaum*, Angew. Chem. *78*, 953 (1966); Angew. Chem. internat. Edit. *5*, 933 (1966).

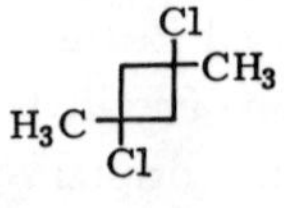

(457)

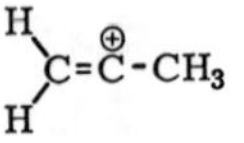

(458) 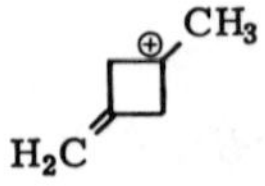 *(459)*

lene by the [$_{\pi}2_s + _{\pi}2_a$] route to give *(459)*, which reacts with chloride ion and hydrogen chloride to afford the observed products. Further, the very useful, but surprising — indeed hitherto baffling — synthesis of 1,2-dichloro-1,2,3,4-tetramethylcyclobutene *(460)*, by treatment of 2-butyne with chlorine in the presence of boron trifluoride[236, 237], is now readily explicable: [$_{\pi}2_s + _{\pi}2_a$] combination of 2-

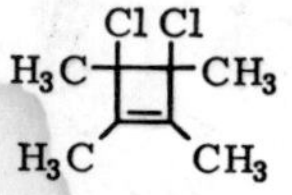

(460)

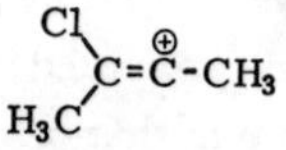

(461) 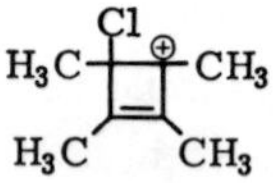 *(462)*

butyne with the β-chlorovinylium ion *(461)* leads directly to the cyclobutenyl cation *(462)*, whose discharge by chloride ion yields *(460)*. These striking examples leave no doubt that vinylium ions possess in high degree the predicted capacity for concerted combination with $_{\pi}2_s$ systems.

And now we can answer emphatically in the affirmative the inquiry which launched the discussion just concluded. Ketenes, in their capacity as vinylium ylides *(453)*, are in fact ideally constituted to play an antarafacial role in reaction with $_{\pi}2_s$ sys-

---

[236] *R. Criegee* and *A. Moschel*, Chem. Ber. *92*, 2181 (1959).

[237] *I. V. Smirnov-Zamkov*, Doklady Akad. Nauk SSSR *83*, 869 (1952); *I. V. Smirnov-Zamkov* and *N. A. Kostromina*, Ukr. Khim. Zhur. *21*, 233 (1953).

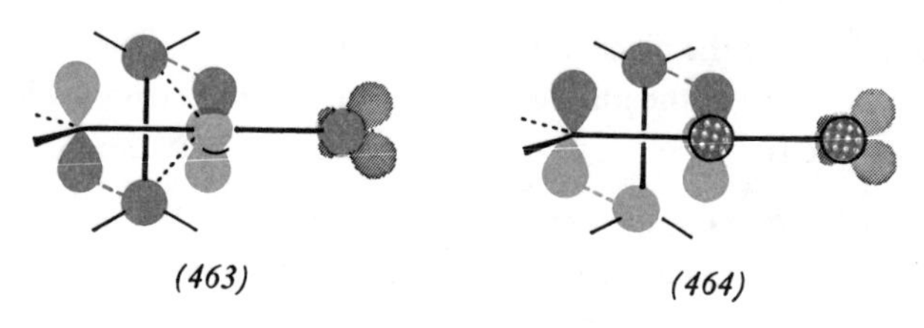

tems, and it is that symmetry-allowed role which they exhibit in the many observed concerted combinations with olefinic systems. Of course, ketenes are not simple vinylium ions, but detailed analysis reveals that the circumstances are not significantly altered in the more complicated case. Thus, for ketenes the relevant orbital diagrams are *(463)* and *(464)*. In this case the place of the vacant p orbital of the simple vinylium ion is taken by the unoccupied $\pi^*_{C=O}$ orbital of the ketene molecule [*cf.* *(463)*]; it is the exceptionally low-lying position of this orbital which accounts in molecular orbital terms for the high electrophilic reactivity of ketenes. As in the simpler case, there is no net change in bonding through interaction of the extra π system with the π^* orbital of the ${}_{\pi}2_s$ component [*cf.* *(456)* and *(464)*]. Finally, it should be pointed out explicitly that interaction of the $\pi_{C=C}$ system of the ketene molecule with the antisymmetric lone pair orbital at the oxygen atom is essentially irrelevant to the matter at hand. This interaction, which is small in terms of energy in any event, creates three new allylic orbitals (Figure 45). Of these, two are occu-

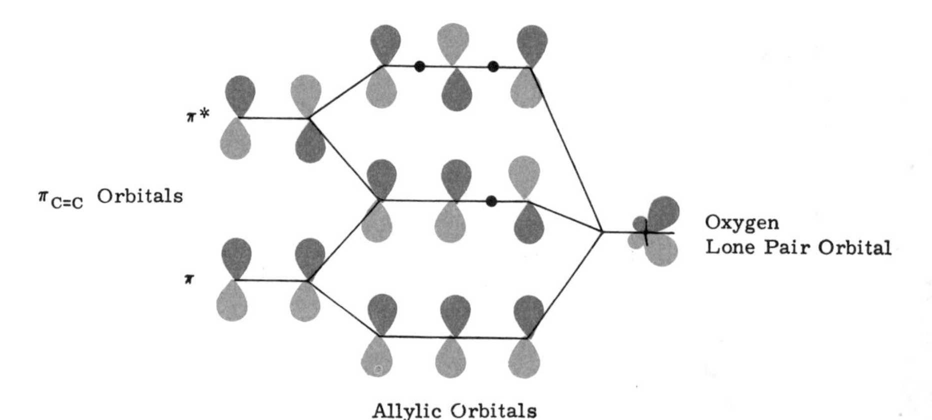

Figure 45. Allylic orbitals in the ketene molecule. — Note that as a consequence of the relative stability of the occupied atomic oxygen orbital, the node in the middle allylic orbital is not at the central atom, as it would be in a three-carbon system, and must be located between carbon and oxygen.

pied; the lower simply takes the place of the usual two-orbital π system [*cf.* *(464)* and *(456)*] and the electrons in the middle orbital return to their original lone pair status as reaction proceeds.

In the dimerization of ketenes — also regarded as concerted[238] — one of the ketene molecules exercises its unique potentiality as a $_{\pi}2_a$ component, while the other acts in normal $_{\pi}2_s$ fashion. In that connection it should be noted that the participation of a ketene as a $_{\pi}2_s$ component is not facilitated by the presence of the orthogonal $\pi_{C=O}$ system, nor is the interaction with the lone pair at oxygen expected to have a major effect. Thus, in participating in cycloaddition reactions as $_{\pi}2_s$ components, ketenes should not differ markedly from normal alkyl ethylenes. Consequently, in the reactions of ketenes with dienes, the normal symmetry-allowed $[_{\pi}4_s+_{\pi}2_s]$ reaction is not expected to be a ready one. Further, the $[_{\pi}4_s+_{\pi}2_a]$ reaction is of course symmetry-forbidden, and the presence of an orthogonal vacant *p* orbital, or low-lying $\pi^*_{C=O}$ system, cannot facilitate the $[_{\pi}4_a+_{\pi}2_a]$ process. Taken together, these circumstances account for the formation of cyclobutanones to the exclusion of cyclohexenones when ketenes react with simple dienes.

---

[238] *R. Huisgen* and *P. Otto*, J. Amer. chem. Soc. *90*, 5342 (1968).

11. Generalized Selection Rules for Pericyclic Reactions

In our development of the theme of orbital symmetry control of concerted chemical changes, we have laid the basis for a general consideration of all *pericyclic* reactions — that is, reactions in which all first-order changes in bonding relationships take place in concert on a closed curve. Thus, we have discussed in detail the special characteristic features of several clearly differentiable reaction types. But we have also made it clear that the different classes have much in common. In particular, we have pointed out that electrocyclic reactions and sigmatropic changes can as well be treated as concerted intramolecular cycloaddition processes. Now, we develop further the important generalization that all pericyclic reactions may be categorized as concerted cycloaddition processes, and must obey the selection rules for such changes.

The selection rules for two-component cycloadditions [*cf.* Figure 22, Section 6] may easily be generalized by induction to include any number of components. In that way we derive the following general rule for all pericyclic changes:

A ground-state pericyclic change is symmetry-allowed when the total number of $(4q+2)_s$ *and* $(4r)_a$ *components is odd.*

Obviously, if the designated total is *even*, the reaction is symmetry-forbidden. Odd-electron systems generally conform to the pattern for even-numbered systems containing one more electron, and there is in general, corresponding to every ground state reaction, a related excited-state process for which the rule is simply reversed.

In order to complete the generalized description of pericyclic reactions, it is necessary at this point to introduce one further, and final, convention. Clearly, when q or $r = 0$, a $4r$ component necessarily represents a single atomic orbital, and a $4q+2$ component may do so. That is to say, such a single orbital component may be vacant or occupied. Like any other component, single orbital components may participate in either a suprafacial or an antarafacial manner, and they require a special designation — a presubscript ω — consistent with the characterization of σ and π components, thus:

$${}_{\omega}0_s \quad {}_{\omega}0_a \quad {}_{\omega}2_s \quad {}_{\omega}2_a$$

With the tools now at hand, it is a simple matter to classify any pericyclic reaction, and to determine whether it is symmetry-allowed. We shall now recapitulate the examples mentioned earlier, and exemplify the procedure in some new instances. Note that the dashed lines in all of the analyses represent bonding interactions in the transition state.

a) The conrotatory electrocyclic opening of cyclobutenes [*cf.* Figures 23 and 24, Section 6] may be regarded either as a $[_\pi 2_s + _\sigma 2_a]$ or a $[_\pi 2_a + _\sigma 2_s]$ process, and the general rule shows at once that it is symmetry-allowed.

b) The suprafacial [1,3] sigmatropic shift with inversion at the migrating center [*cf.* Figure 34, Section 7] may also be regarded as a symmetry-allowed $[_\pi 2_s + _\sigma 2_a]$ or a $[_\pi 2_a + _\sigma 2_s]$ change.

c) The disrotatory electrocyclic transformation of a cyclopropyl cation to an allyl cation may be regarded either as an $[_\omega 0_s + _\sigma 2_s]$ *(465)* or as an $[_\omega 0_a + _\sigma 2_a]$ *(466)* pro-

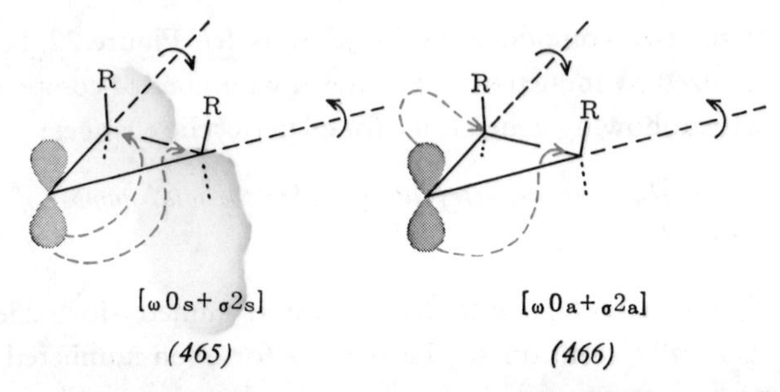

$[_\omega 0_s + _\sigma 2_s]$ *(465)* $[_\omega 0_a + _\sigma 2_a]$ *(466)*

cess, and is symmetry-allowed. Note that in a formal sense, when a single ω component is involved in a pericyclic reaction, the ω orbital moves to a new position, and one of the dashed lines used in the analysis vanishes in the product.

d) The [1,2] sigmatropic shift within a cation cannot take place with inversion of the migrating group, since the $[_\omega 0_a + _\sigma 2_s]$ *(467)* and $[_\omega 0_s + _\sigma 2_a]$ *(468)* processes are symmetry-forbidden: Note that antarafacial bonding interaction between two

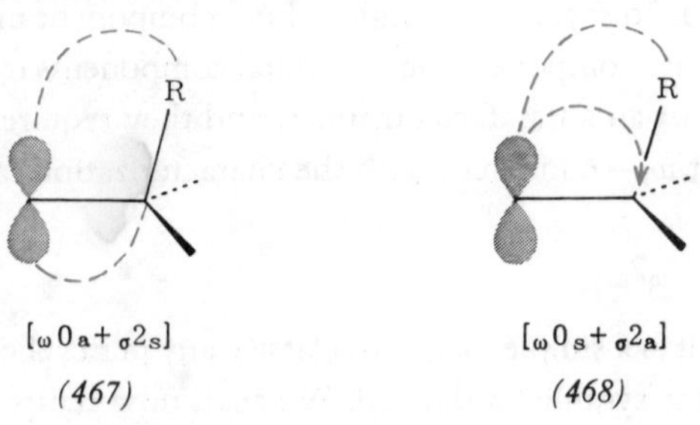

$[_\omega 0_a + _\sigma 2_s]$ *(467)* $[_\omega 0_s + _\sigma 2_a]$ *(468)*

atoms joined by a non-participating bond and bridged by a single atom is physically impossible. Thus, the $[_{\omega}0_s + _{\sigma}2_s]$ processes depicted in *(469)* and *(470)* both repre-

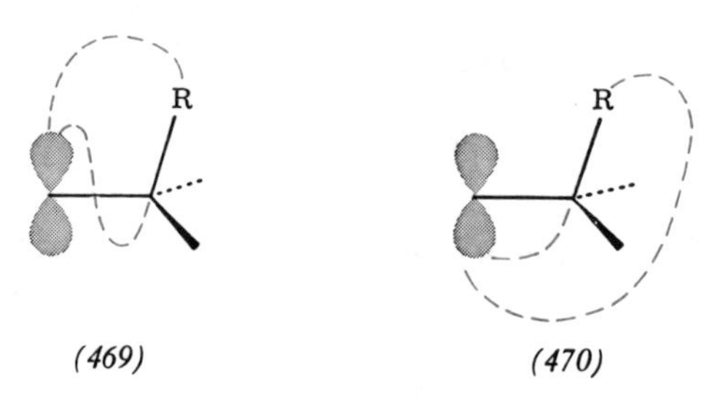

(469) *(470)*

sent the symmetry-allowed but physically unrealizable antarafacial [1,2] sigmatropic shift with inversion at the migrating center.

e) The suprafacial [1,4] sigmatropic shift with inversion at the migrating center may be regarded, *i.a.*, as a symmetry-allowed $[_{\pi}2_a + _{\omega}0_a + _{\sigma}2_a]$ process *(471)*. All of

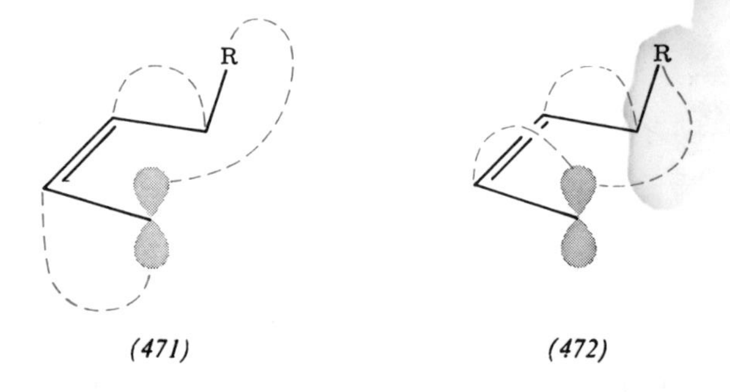

(471) *(472)*

the alternative formal classifications *(473)*, each of which represents a symmetry-allowed reaction, require inversion at R, and no allowed suprafacial process exists in which R migrates with retention; for example, the $[_{\pi}2_s + _{\omega}0_s + _{\sigma}2_s]$ pericyclic reaction depicted in *(472)* is obviously symmetry-forbidden.

$$\begin{aligned} &[_{\pi}2_s + {}_{\omega}0_a + {}_{\sigma}2_s] \\ &[_{\pi}2_s + {}_{\omega}0_s + {}_{\sigma}2_a] \\ &[_{\pi}2_a + {}_{\omega}0_s + {}_{\sigma}2_s] \end{aligned} \qquad (473)$$

f) The non-linear cheletropic extrusion of sulfur dioxide from an episulfone [*cf.* *(418)*, Section 10.1] is a symmetry-allowed $[_{\sigma}2_s + _{\sigma}2_a]$ reaction *(474)*, and the reverse process would be a $[_{\pi}2_s + _{\omega}2_a]$ change *(475)*.

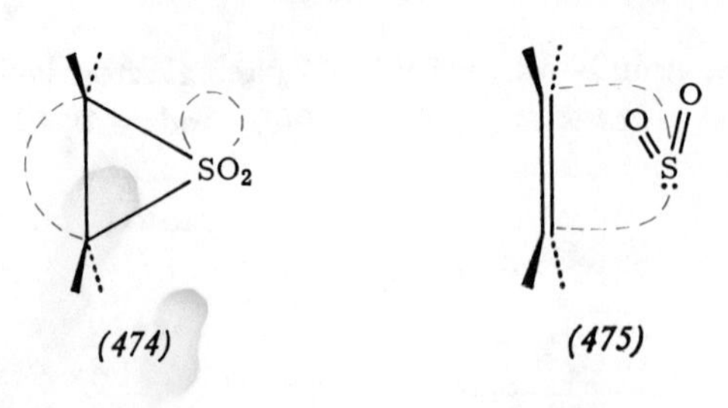

(474) *(475)*

g) The Diels-Alder reaction is obviously a symmetry-allowed $[_{\pi}4_s+_{\pi}2_s]$ change, but it could be analyzed as an allowed $[_{\omega}0_s+_{\omega}2_s+_{\omega}0_a+_{\omega}2_a+_{\omega}0_s+_{\omega}2_s]$ process *(476)*.

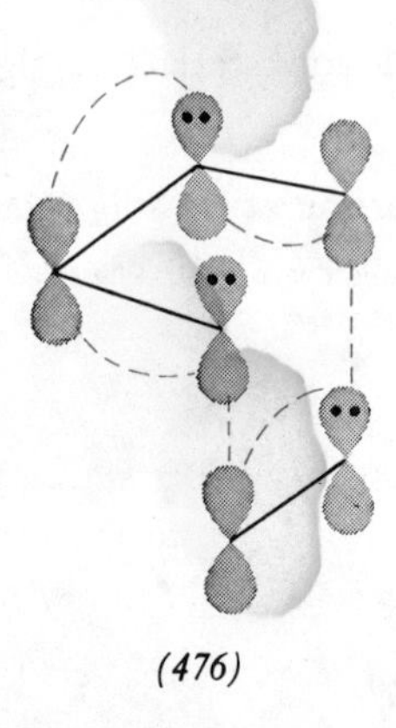

(476)

The example is included partly in a sportive spirit, but it does have the serious purposes of indicating that (i) the participation of any number and kind of ω components in a pericyclic reaction presents no occasion for difficulty in analyzing the process in terms of the general rule, and (ii) polarization of the bonds involved in a pericyclic reaction in no way alters the orbital symmetry control to which it is subject.

It should be emphasized that use of the general selection rule for pericyclic reactions must always be subject to the following precautions.

a) The geometry of each individual case must be examined on its merits in order to ascertain whether the process is physically realizable. We have already exemplified, in d) above, and in our discussion of the rearrangement of the ester of Feist's acid [*cf.* *(331)*, Section 7.1], situations in which formally symmetry-allowed processes are geometrically impossible.

b) In cases in which the reacting components are directly joined by non-participating bonds, forced ancillary off-circuit antibonding interactions may render an otherwise allowed reaction symmetry-forbidden. Thus, prismane cannot be converted to three isolated, non-interacting, ground state ethylene moieties by a $[_{\sigma}2_s+_{\sigma}2_a+_{\sigma}2_a]$ process. The geometric constraints imposed by the non-participant σ framework require that the ethylene moieties produced in such a reaction interact strongly in an antibonding way, and the orbital symmetry-imposed electronic configuration of the array is in fact that of a doubly excited state of the actual product benzene [*cf.* Section 6.4]. Any case in which similar circumstances may be suspected should be subjected individually to complete analysis in the light of the principle of orbital symmetry conservation.

12. Violations

There are none!

Nor can violations be expected of so fundamental a principle of maximum bonding. All the more is it then important to give consideration to some reactions which might appear on casual inspection to contravene orbital symmetry conservation.

First, it should be emphasized explicitly here that a given symmetry-allowed concerted reaction need not necessarily represent the manner in which reacting molecules will actually comport themselves. Quite possibly another path, involving reactive *intermediates* of relatively low energy may be followed. The other side of this coin is that molecules may in some cases combine, or decompose, to give the very products which would result from symmetry-forbidden concerted processes. In these cases it is clear that the actual reaction path is a non-concerted one, involving discrete intermediates; indeed, the principle of orbital symmetry conservation is very useful in defining conditions in which such processes may be expected. Such are often the circumstances in the combination of ethylenes to give cyclobutanes. The direct, concerted $[_{\pi}2_s+_{\pi}2_s]$ combination is of course symmetry-forbidden, and the allowed $[_{\pi}2_s+_{\pi}2_a]$ reaction is realizable only in special circumstances. Nevertheless, there are a substantial number of instances, lacking the special prerequisites for the $[_{\pi}2_s+_{\pi}2_a]$ reaction, in which the combination of two ethylenic molecules leads to a cyclobutane[239]. And in such cases, the available evidence is compelling in favor of the view that discrete diradical or ionic intermediates are involved[240]; further, it is of obvious significance that such reactions are observed only when the reactant olefins are ornamented by substituents so constituted as to provide effective stabilization of the requisite intermediate.

It remains only to address a few final comments to the matter of symmetry-forbidden reactions which may occur in a concerted fashion. In principle, of course, given sufficient energy, and means of constraining molecules in energy-rich configurations, we could bring about any symmetry-forbidden reaction. A very powerful Maxwell demon could make a molecule of cyclobutane by grasping a molecule of ethylene in each hand, and — not without an impressive exhibition of his strength — forcing the two together, face-to-face. He could seize a molecule of cyclobutene at its methylene groups, and tear it open in a disrotatory fashion to obtain butadiene. But without such demoniac intervention, molecules will very rarely comport themselves in like fashion — since reaction paths with lower energy surfaces than those associated with symmetry-forbidden processes will almost always be available. None the less, the possibility remains that molecules may be constructed within which no other such alternatives exist; then, provided that the molecule be sufficiently energized, the symmetry-forbidden path would necessarily be followed. Derivatives of *cis*-bicyclo[2.2.0]hexadiene *(477)* quite possibly exem-

(477)

plify such a process. These substances do undergo conversion to the corresponding benzenoid isomers, with an activation energy in the neighborhood of 37 kcal/mole[241]. The simple disrotatory cleavage of the central bond is symmetry-forbidden, but *(477)* is already strained to the extent of *ca.* sixty kilocalories rela-

---

[239] An excellent review is available: *J. D. Roberts* and *C. M. Sharts*, Organic Reactions *12*, 1 (1962).

[240] *P. D. Bartlett, L. K. Montgomery*, and *B. Seidel*, J. Amer. chem. Soc. *86*, 616 (1964); *L. K. Montgomery, K. Schueller*, and *P. D. Bartlett*, ibid. *86*, 622 (1964); *P. D. Bartlett* and *L. K. Montgomery*, ibid. *86*, 628 (1964); *P. D. Bartlett*, Science *159*, 833 (1968).

[241] *J. F. M. Oth*, Angew. Chem. *80*, 633 (1968); Angew. Chem. internat. Edit. *7*, 646 (1968); Rec. Trav. Chim. Pays-Bas *87*, 1185 (1968); *H. C. Volger* and *H. Hogeveen*, Rec. Trav. Chim. Pays-Bas *86*, 830 (1967).

tive to benzene[242], and the addition of a further thirty-seven kilocalories may sufficiently energize the molecule as to overcome the prohibition against disrotatory cleavage. But even in this case, the possibility is not entirely excluded that events may take another course, and that the transformation may proceed through a discrete intermediate. For example, if the breaking of the central bond of *(477)* is accompanied by a quasi-conrotatory skewing distortion [*cf.* *(478)*] — a boat→twist-boat transformation — the result *(479)* is an array containing two skewed allyl radical

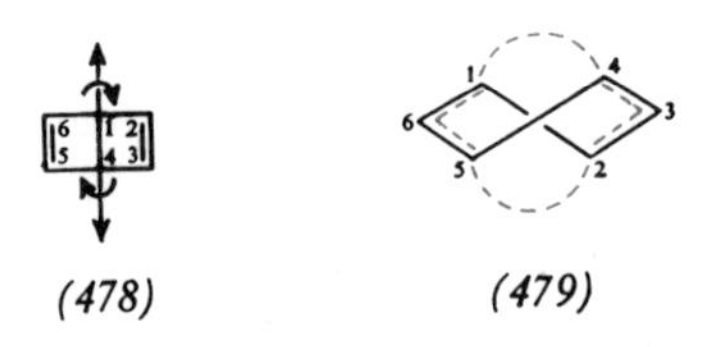

(478) *(479)*

systems [C(5)-C(6)-C(1) and C(2)-C(3)-C(4)], joined orthogonally at their termini. Were all bonding between C-1 and C-4 to be lost concurrently with these motions, the related symmetry-derived antibonding components in the transition state for conversion to benzene would vanish, and the process could continue to completion by appropriate symmetry-allowed combinations of the isolated allyl radicals. Even if some 1,4 bonding is retained — and of course, were that the case, some 2,5 bonding would be generated — the corresponding symmetry-derived antibonding components in the transition state *(479)* will be smaller than those in the transition state for the simple symmetry-forbidden disrotatory cleavage; it is of special interest that if *(479)* should represent the transition state for the conversion of bicyclohexadienes to aromatic compounds, the former should be susceptible of isomerization in the sense *(480)*→*(481)*→*(482)*.

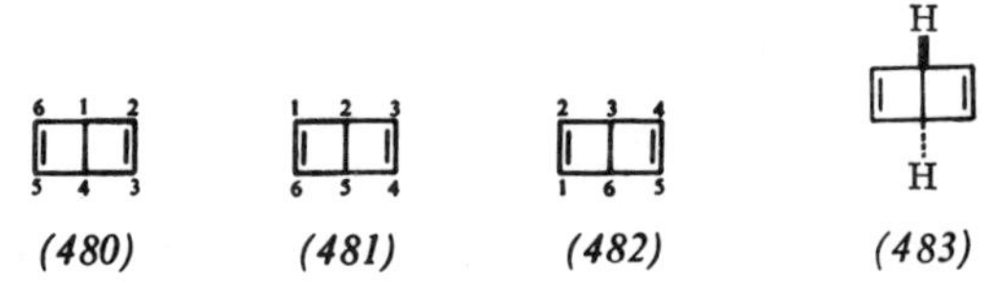

(480) *(481)* *(482)* *(483)*

Finally, we emphasize the fact that *cis*-bicyclo[2.2.0]hexadienes owe their remarkable durability to symmetry-imposed barriers, and, to introduce a light note, suggest that our fellow chemists estimate how long the isomer *(483)* — put together, perhaps, by our versatile demon — would survive.

---

[242] *W. Schäfer*, Angew. Chem. *78*, 716 (1966); Angew. Chem. internat. Edit. *5*, 669 (1966).

13. Other Theoretical Work

In view of the fact that the principle of conservation of orbital symmetry has been found to be an exceptionally powerful predictive and interpretive tool, it is not surprising that a number of alternative theoretical approaches have been set forth. In all cases the central conclusions, of necessity, agree in all respects with our own. Reference should be made to the interesting contributions of *Fukui*[243], *Salem*[244], *Zimmerman*[245], *Dewar*[246], *Oosterhoff* and *van der Lugt*[247], and, of course, *Longuet-Higgins* and *Abrahamson*[248].

It is also appropriate here to mention some earlier studies relevant to and antecedent to our own. Correlation diagrams for the simplest chemical reactions are no novelty; some interesting applications are presented by *Laidler*[249]. Such diagrams were extensively used by *Griffing*[250], but were incorrectly interpreted. In an important paper by *Herzfeld*[251] it was shown that for a set of $2p$ orbitals the number of nodes in a molecular system is conserved as atoms are brought near each other, and that when the orbitals are alike no two states of different energy can have the same number of nodes. *Bader*[252] presented a novel and interesting discussion of some elementary reactions.

We have already mentioned the suggestion of *Oosterhoff*[253] that orbital symmetry might play a role in electrocyclic reactions. The advantages of a six-electron transition state were pointed out by *Syrkin*[254]. His papers, and those by *Balaban*[255] and

---

[243] *K. Fukui*, Tetrahedron Letters *1965*, 2009, 2427; Bull. chem. Soc. Japan *39*, 498 (1966); *K. Fukui* and *H. Fujimoto*, Tetrahedron Letters *1966*, 251; Bull. chem. Soc. Japan *39*, 2116 (1966); *40*, 2018 (1967).

[244] *L. Salem*, J. Amer. chem. Soc. *90*, 543, 553 (1968).

[245] *H. E. Zimmerman*, J. Amer. chem. Soc. *88*, 1563, 1566 (1966).

[246] *M. J. S. Dewar*, Tetrahedron, Suppl. 8, p. 75 (1966); Aromaticity. Special Publication of The Chemical Society No. 21, London 1967, p. 177.

[247] *W. Th. A. M. van der Lugt* and *L. J. Oosterhoff*, Chem. Commun. *1968*, 1235.

[248] *H. C. Longuet-Higgins* and *E. W. Abrahamson*, J. Amer. chem. Soc. *87*, 2045 (1965).

[249] *K. J. Laidler:* The Chemical Kinetics of Excited States. Oxford University Press, Oxford 1955.

[250] *V. Griffing*, J. chem. Physics *25*, 1015 (1955); J. phys. Chem. *61*, 11 (1957).

[251] *K. F. Herzfeld*, Z. Naturforsch. *3a*, 457 (1948); Rev. mod. Physics *41*, 527 (1949).

[252] *R. F. W. Bader*, Canad. J. Chem. *40*, 1164 (1962). *Cf.* also *R. G. Pearson*, J. Amer. chem. Soc. *91*, 1252 (1969), and *L. Salem*, Chem. Phys. Letters *3*, 99 (1969).

[253] *L. J. Oosterhoff*, quoted in *E. Havinga* and *J. L. M. A. Schlatmann*, Tetrahedron *16*, 151 (1961).

[254] *Ya. K. Syrkin*, Izv. Akad. Nauk SSSR *1959*, 238, 389, 401, 600.

[255] *A. T. Balaban*, Rev. Roum. Chimie *11*, 1097 (1966); *12*, 875 (1967).

Mathieu and *Valls*[256] are a rich source of references for possible concerted reactions.

Numerous reviews summarizing various aspects of the application of the principle of conservation of orbital symmetry have already appeared[257].

The frontier-electron theory was developed by *Fukui*, who has continued an intensive exploration of various methods of calculation pertinent to problems of chemical reactivity[258]. In some of our work we have utilized extended Hückel calculations[259] to confirm qualitative conclusions. The virtue of these approximate calculations lies in the fact that they may be applied to σ as well as π systems. The method was originally developed in collaboration with *Lipscomb* and owes much to previous work of *Longuet-Higgins* on boron hydrides.

Of special interest are several recent applications of orbital symmetry conservation in novel directions. The derivation of selection rules for the isomerization and substitution reactions of transition metal complexes[260], the suggestion that orbital symmetry is determinative in chemiluminescent reactions[261], and the analysis of the role of orbital symmetry in excited state energy transfer[262] merit particular attention.

---

[256] *J. Mathieu* and *J. Valls*, Bull. Soc. chim. France *1957*, 1509.
[257] *Inter alia*, *J. J. Vollmer* and *K. L. Servis*, J. Chem. Educat. *45*, 214 (1968); *S. I. Miller* in *V. Gold:* Advances in Physical Organic Chemistry. Academic Press, New York 1968, Vol. 6, p. 185; *G. B. Gill*, Quart. Rev. *22*, 338 (1968); *D. Seebach*, Fortschr. chem. Forsch. *11*, 177 (1969); *P. Millie*, Bull. Soc. chim. France *1966*, 4031; *M. Orchin* and *H. H. Jaffé:* The Importance of Antibonding Orbitals. Houghton-Mifflin, Boston 1967; *E. M. Kosower:* An Introduction to Physical Organic Chemistry. Wiley, New York 1968; *O. Červinka* and *O. Kříž*, Chem. Listy *61*, 1036 (1967); *62*, 321 (1968); *J. P. M. Houbiers*, Chem. Weekblad *62*, 61 (1966); *C. Hörig*, Z. Chem. *7*, 298 (1967); *M. J. S. Dewar:* The Molecular Orbital Theory of Organic Chemistry. McGraw-Hill, New York 1969.
[258] *K. Fukui* in *P.-O. Löwdin* and *B. Pullman:* Molecular Orbitals in Chemistry, Physics and Biology. Academic Press, New York 1964, p. 513; *K. Fukui* in *O. Sinanoğlu:* Modern Quantum Chemistry. Academic Press, New York 1965, p. 49.
[259] *R. Hoffmann* and *W. N. Lipscomb*, J. chem. Physics *36*, 2179 (1962); *37*, 2872 (1962); *R. Hoffmann*, ibid. *39*, 1397 (1963); *40*, 2474, 2480, 2745 (1964); Tetrahedron *22*, 521, 539 (1966).
[260] *D. R. Eaton*, J. Amer. chem. Soc. *90*, 4272 (1968).
[261] *F. McCapra*, Chem. Commun. *1968*, 155.
[262] *E. F. Ullman*, J. Amer. chem. Soc. *90*, 4158 (1968).

14. Conclusion

We have now explicated the principle of conservation of orbital symmetry, and exemplified its use. It may perhaps be appropriate here to emphasize that the central content of the principle lies in the incontrovertible proposition that a chemical reaction will proceed the more readily, the more *bonding* may be maintained throughout the transformation. Consequently, we cannot doubt that the principle will endure, whatever the language in which it may be couched, or whatever greater precision may be developed in its application and extension.

Nor can it be doubted that extension will be made — both of a fundamental and of a detailed nature. Since every elementary step in *any* chemical reaction is a concerted process, correlative ideas must be applicable to *all* reactions. The exploration of applications in inorganic chemistry has barely been begun. And within the realm already delineated, only the imagination sets limits upon the number and variety of new and fascinating molecules which may be designed to favor reactions as yet undiscovered, but now known to be feasible.